SpringerBriefs in Mathematics

Series Editors

Nicola Bellomo, Torino, Italy

Michele Benzi, Pisa, Italy

Palle Jorgensen, Iowa City, USA

Roderick Melnik, Waterloo, Canada

Otmar Scherzer, Linz, Austria

Benjamin Steinberg, New York, NY, USA

Lothar Reichel, Kent, USA

Yuri Tschinkel, New York, NY, USA

George Yin, Detroit, USA

Ping Zhang, Kalamazoo, MI, USA

SpringerBriefs present concise summaries of cutting-edge research and practical applications across a wide spectrum of fields. Featuring compact volumes of 50 to 125 pages, the series covers a range of content from professional to academic. Briefs are characterized by fast, global electronic dissemination, standard publishing contracts, standardized manuscript preparation and formatting guidelines, and expedited production schedules.

Typical topics might include:

- A timely report of state-of-the art techniques
- A bridge between new research results, as published in journal articles, and a contextual literature review
- A snapshot of a hot or emerging topic
- An in-depth case study
- A presentation of core concepts that students must understand in order to make independent contributions

SpringerBriefs in Mathematics showcases expositions in all areas of mathematics and applied mathematics. Manuscripts presenting new results or a single new result in a classical field, new field, or an emerging topic, applications, or bridges between new results and already published works, are encouraged. The series is intended for mathematicians and applied mathematicians. All works are peer-reviewed to meet the highest standards of scientific literature.

Titles from this series are indexed by Scopus, Web of Science, Mathematical Reviews, and zbMATH.

Thai Son Doan · Peter E. Kloeden ·
Hoang The Tuan

Attractors of Caputo Fractional Differential Equations

 Springer

Thai Son Doan
Institute of Mathematics
Vietnam Academy of Science
and Technology
Hanoi, Vietnam

Peter E. Kloeden
Universität Tübingen
Fachbereich Mathematik
Tübingen, Germany

Hoang The Tuan
Institute of Mathematics
Vietnam Academy of Science
and Technology
Hanoi, Vietnam

ISSN 2191-8198   ISSN 2191-8201 (electronic)
SpringerBriefs in Mathematics
ISBN 978-3-032-05510-1   ISBN 978-3-032-05511-8 (eBook)
https://doi.org/10.1007/978-3-032-05511-8

This Springer imprint is published by the registered company Springer Nature Switzerland AG
The registered company address is: Gewerbestrasse 11, 6330 Cham, Switzerland

If disposing of this product, please recycle the paper.

(Thai Son Doan) my wife and children

*(Peter E. Kloeden) Karin Wahl-Kloeden
1954–2016*

*(Hoang The Tuan) my wife, Nguyen Phuong
Lien and children, Hoang Ngoc Gia Han and
Hoang The Gia Phat*

Preface

The theory of fractional analysis and fractional differential equations is vast and has deep historical roots. The introduction of Caputo fractional differential equations (FDEs) in the 1960s allowed initial value problems to be handled more naturally and the asymptotic behaviour of models based on them to be investigated. More recently, mathematically defined dynamical systems generated by Caputo FDEs and their attractors have been introduced.

Dissipative Caputo FDEs have vector fields which satisfy a dissipativity property. For ordinary differential equations (ODEs) it follows from such a property that an absorbing set exists which contains all of the long-term dynamical behaviour of the system such as the existence of an attractor. The situation is more complicated for Caputo FDEs, since these are essentially integral equations and the dissipative inequalities cannot be so easily exploited. Moreover, such integral equations are essentially nonautonomous due to the form of the kernel in the integral equations, even when the vector field is "autonomous", i.e., does not depend explicitly on time.

In this book we focus on dissipative Caputo FDEs with an autonomous vector field, although many of the results considered here also hold in the nonautonomous case. (These can often be found in the cited literature). This is primarily to simplify the technical nature of many proofs and to allow us to focus on the essential differences from the ODE case that arise due to use of fractional calculus. Another reason is to avoid the errors that have appeared more often than desired in the published literature due to a number of causes, in particular, it seems, the misleading use of intuition based on classical calculus in the fractional context.

The book is based on recent results of the three coauthors in various combinations with each other and with their other coauthors, in particular Nguyen Dinh Cong and Hieu Trinh. The main aim is to develop and present a theory of dynamical systems and their attractors for Caputo FDEs.

We make repeated use of the monographs of Diethelm [1] and Kilbas, Srivastava & Trujillo [2] for many background results on fractional analysis and FDEs, which will often be used without direct reference. The reader can find the original sources of these results there. Also many of the results that we take from the literature often have their own precedents for which we refer readers to the discussion in our primary references for more detail.

The authors are very grateful to Prof. Nguyen Dinh Cong for his encouragement and helpful advice during the preparation of this book. They also thank Cui Hongyong, Sun Wenlong and Duc Thai Ha for carefully reading parts of the manuscript and their helpful comments.

Financial support from a Simons Foundation grant to the Institute of Mathematics of the Vietnam Academy of Science and Technology is gratefully acknowledged. TSD was funded by Vietnam Ministry of Education and Training under Grant Number B2023-CTT-07. PEK thanks the Vietnam Academy of Science and Technology in Hanoi for its hospitality. PEK was also funded partially by Ministerio de Ciencia e Innovación (Spain) and FEDER (European Community) under grant PID2021-122991NB-C21 and the Chinese NSF grant 12371242.

Hanoi, Vietnam   Thai Son Doan
Tübingen, Germany   Peter E. Kloeden
Hanoi, Vietnam   Hoang The Tuan
July 2025

References

1. Diethelm, K.: The Analysis of Fractional Differential Equations, Springer Lecture Notes in Mathematics, Vol. 2004, Springer, Heidelberg (2010)
2. Kilbas, A.A., Srivastava, H. M., Trujillo, J. J.: Theory and applications of fractional differential equations, North-Holland Mathematics Studies 204, Elsevier Science B.V., Amsterdam, MR2218073, (2006)

Contents

Chapter 1
Introduction

Abstract Several fundamental concepts, definitions, and preparatory materials that will be used throughout the book are introduced, in particular, the Riemann–Liouville fractional integral, the Caputo fractional derivative, Mittag-Leffler functions, and their essential properties.

Keywords Riemann-Liouville fractional integral · Caputo fractional derivative · Mittag-Leffler functions · Variation of constants formula

This monograph focuses on autonomous Caputo fractional differential equations (FDE) of order $\alpha \in (0, 1)$ in $\mathbb{R}^d$ of the form

$$^{C}D_{0+}^{\alpha} x(t) = g(x(t)) \tag{1.1}$$

where $g : \mathbb{R}^d \to \mathbb{R}^d$ is globally Lipschitz continuous or locally Lipschitz continuous and also satisfies a dissipativity condition such as

$$\langle x, g(x) \rangle \leq a - b \|x\|^2, \tag{1.2}$$

where $a, b > 0$. The aim is to investigate the long time, i.e., asymptotic behaviour of their solutions, in particular, the nature of their limits sets and attractors.

Fractional differential equations have a very long history with an extensive literature and with many applications. The Caputo version of the fractional derivative used here is more recent and is more amenable to a dynamical system investigation than the classical versions such as the Riemann-Liouville fractional derivative.

The fractional derivative does not have the convenient properties and geometric association that the ordinary derivative has, e.g., a negative fractional derivative does not necessarily imply that the function is decreasing. In particular, the rules of differential calculus do not apply. This makes the analysis of Caputo FDEs more complicated than for corresponding ordinary differential equations (ODEs).

A continuous function $x : [0, T] \to \mathbb{R}^d$ is called a *solution* of (1.1) if it satisfies (1.1) for all $t \in [0, T]$ and the value $x(0)$ is called the initial value of the solution

$x(\cdot)$. The initial value problem of the Caputo FDE (1.1) is equivalent to a Volterra integral equation. This greatly facilitates the investigation of Caputo FDEs.

Lemma 1.1 *A continuous function $x : [0, T] \to \mathbb{R}^d$ is a solution of the FDE (1.1) with the initial value condition $x(0) = x_0$ if and only if it is a solution of the Volterra integral equation of the second kind*

$$x(t) = x_0 + \frac{1}{\Gamma(\alpha)} \int_0^t (t - s)^{\alpha-1} g(x(s)) ds, \ \forall t \in [0, T], \tag{1.3}$$

where $\Gamma(\alpha) := \int_0^\infty t^{\alpha-1} e^{-t} dt$ is the Gamma function.

The proof in one-dimensional case can be found in [1, Lemma 6.2, p. 86] and [2, Theorem 3.24, p. 199]. The multi-dimensional case is just a componentwise application of the one-dimensional case. Note that the mapping $\alpha \mapsto \Gamma(\alpha)$ is continuous.

For notational compactness, the integral equation (1.3) will often be written

$$x(t) = x_0 + \int_0^t a_\alpha(t, s) g(x(s)) ds, \ t \in [0, T], \tag{1.4}$$

where

$$a_\alpha(t, s) := \frac{1}{\Gamma(\alpha)} (t - s)^{\alpha-1}.$$

The subscript α on $a_\alpha(t, s)$ will be sometimes omitted when it is clear from the context.

The integral equation (1.3) is a special case of the more general Volterra integral equation of the second kind (see, e.g., [3])

$$x(t) = f(t) + \frac{1}{\Gamma(\alpha)} \int_0^t (t - s)^{\alpha-1} g(x(s)) ds, \ \forall t \in [0, T], \tag{1.5}$$

where $f : \mathbb{R}_{\geq 0} \to \mathbb{R}^d$ is a continuous function. In the case of an Caputo FDE (1.1), $f(t) \equiv f(0) = x_0$.

The following integral will be used repeatedly in the sequel.

$$\int_{\tau_1}^{\tau_2} (t - s)^{\alpha-1} ds = \frac{1}{\alpha} \left((t - \tau_1)^\alpha - (t - \tau_2)^\alpha \right), \quad 0 \leq \tau_1 \leq \tau_2 \leq t. \tag{1.6}$$

1.1 Some Notation

For $T > 0$, let $C([0, T], \mathbb{R}^d)$ denote the space of continuous functions $x : [0, T] \to \mathbb{R}^d$. It is well known that $\left(C_\infty(\mathbb{R}^d), \| \cdot \|_\infty \right)$ is a Banach space with the norm

$$\|\xi\|_\infty := \sup_{t \in \mathbb{R}_{\geq 0}} \|\xi(t)\| < \infty.$$

For $\alpha \in (0, 1]$, let $\mathcal{H}^\alpha([0, T], \mathbb{R}^d)$ be the Hölder space consisting of functions $v \in C([0, T], \mathbb{R}^d)$ such that

$$\|v\|_{\mathcal{H}^\alpha} := \max_{0 \leq t \leq T} \|v(t)\| + \sup_{0 \leq s < t \leq T} \frac{\|v(t) - v(s)\|}{(t - s)^\alpha} < \infty,$$

and let $\mathcal{H}_0^\alpha([0, T], \mathbb{R}^d)$ denote the closed subspace of $\mathcal{H}^\alpha([0, T], \mathbb{R}^d)$ consisting of functions $v \in \mathcal{H}^\alpha([0, T], \mathbb{R}^d)$ such that $v(0) = 0$ and

$$\sup_{0 \leq s < t \leq T, t-s \leq \varepsilon} \frac{\|v(t) - v(s)\|}{(t - s)^\alpha} \to 0 \quad \text{as } \varepsilon \to 0.$$

1.2 Fractional Derivatives and Integrals

Let $x \in C([0, T], \mathbb{R})$ and $\alpha \in (0, 1)$. The Riemann–Liouville integral of order α is defined by

$$I_{0+}^\alpha x(t) := \frac{1}{\Gamma(\alpha)} \int_0^t (t - \tau)^{\alpha-1} x(\tau) \, d\tau \quad \text{for } t \in (0, T],$$

and the corresponding Riemann–Liouville fractional derivative of order α by

$$^R D_{0+}^\alpha x(t) := \left(\frac{d}{dt} I_{0+}^{1-\alpha} x \right)(t) \quad \text{for } t \in (0, T].$$

Then the *Caputo fractional derivative* $^C D_{0+}^\alpha x$ of x is defined by

$$^C D_{0+}^\alpha x(t) :=^R D_{0+}^\alpha (x(t) - x(0)) \quad \text{for } t \in (0, T],$$

see [1, Definition 3.2, pp. 50].

For a d-dimensional vector function $x(t) = (x_1(t), \ldots, x_d(t))^T$, the Caputo fractional derivative is defined component-wise as

$$^C D_{0+}^\alpha x(t) = \left(^C D_{0+}^\alpha x_1(t), \ldots, ^C D_{0+}^\alpha x_d(t) \right)^T.$$

The following result from Vainikko [4, Theorem 5.2, pp. 475] characterises functions that have a Caputo fractional derivative.

Theorem 1.1 *For $\alpha \in (0, 1)$ and $v \in C([0, T]; \mathbb{R}^d)$, the following conditions are equivalent:*

(i) *the fractional derivative* $^C D_{0+}^\alpha v \in C([0, T], \mathbb{R}^d)$ *exists;*

(ii) *a finite limit* $\lim_{t \to 0} \frac{v(t) - v(0)}{t^\alpha} := \gamma$ *exists, and*

$$\sup_{0 < t \leq T} \left\| \int_{\theta t}^{t} (t - \tau)^{-\alpha - 1} (v(t) - v(\tau))\, d\tau \right\| \to 0 \quad \text{as} \ \theta \to 1,$$

i.e., the integral $\int_0^t (t - \tau)^{-\alpha - 1}(v(t) - v(\tau)) d\tau$ *is equi-convergent for any* $t \in (0, T]$;

(iii) v *has the structure* $v - v(0) = t^\alpha \gamma + v_0$, *where* γ *is a constant vector,* $v_0 \in \mathcal{H}_0^\alpha([0, T], \mathbb{R}^d)$, *and* $\int_0^t (t - \tau)^{-\alpha - 1}(v(t) - v(\tau)) d\tau =: w(t)$ *is equi-convergent for every* $t \in (0, T]$ *defining a function* $w \in C((0, T], \mathbb{R}^d)$ *which has a finite limit* $\lim_{t \to 0} w(t) =: w(0)$.

Moreover, if $v \in C([0, T], \mathbb{R}^d)$ *with* $^C D_{0+}^\alpha v \in C([0, T], \mathbb{R}^d)$, *then* $^C D_{0+}^\alpha v(0) = \Gamma(\alpha + 1)\gamma$, *and*

$$^C D_{0+}^\alpha v(t) = \frac{v(t) - v(0)}{\Gamma(1 - \alpha)t^\alpha} + \frac{\alpha}{\Gamma(1 - \alpha)} \int_0^t \frac{v(t) - v(\tau)}{(t - \tau)^{\alpha + 1}} d\tau$$

for $0 < t \leq T$.

Let $I_{0+}^\alpha C([0, T], \mathbb{R}^d)$ denote the space of functions $\varphi : [0, T] \to \mathbb{R}^d$ such that there exists a function $\psi \in C([0, T], \mathbb{R}^d)$ satisfying $\varphi = I_{0+}^\alpha \psi$. Proposition 6.4 of Vainikko [4] says that Caputo fractional derivative $^C D_{0+}^\alpha x(t)$ exists and is continuous if $x \in I_{0+}^\alpha C([0, T], \mathbb{R}^d)$.

The following result from Kilbas et al. [2] provides an equivalent integral equation representation of the Caputo fractional derivative. See also [4, Theorem 5.2, pp. 475].

Lemma 1.2 *Let* $\alpha \in (0, 1)$ *and let* $h : [0, T] \to \mathbb{R}^d$ *be continuous. Then a function* $x : [0, T] \to \mathbb{R}^d$ *is a solution of the fractional integral equation*

$$x(t) = x_0 + \frac{1}{\Gamma(\alpha)} \int_0^t (t - s)^{\alpha - 1} h(s) ds, \ \forall t \in [0, T], \tag{1.7}$$

if and only if x *is a solution of the fractional Initial Value Problem*

$$\begin{cases} ^C D_{0+}^\alpha x(t) &= h(t), \ t \in (0, T], \\ x(0) &= x_0. \end{cases} \tag{1.8}$$

1.3 Mittag-Leffler Functions

The Mittag-Leffler function [1, 2, 5] is essentially a generalisation of the exponential function. It is defined by

$$E_\alpha(t) = \sum_{k=0}^{\infty} \frac{t^k}{\Gamma(k\alpha + 1)}, \quad t \in \mathbb{R},$$

with $\alpha > 0$, which need not be an integer. A two parameter generalisation is defined by

$$E_{\alpha,\beta}(t) = \sum_{k=0}^{\infty} \frac{t^k}{\Gamma(k\alpha + \beta)}, \quad t \in \mathbb{R},$$

with $\alpha, \beta > 0$ and $t \in \mathbb{R}$. Clearly, $E_\alpha(t) = E_{\alpha,1}(t)$.

Theorem 1.2 [6, Lemma 2.1] *For any $\alpha \in (0, 1)$, the Mittag-Leffler function $E_\alpha : \mathbb{R} \to \mathbb{R}$ is strictly monotonically increasing with*

$$\lim_{t \to \infty} E_\alpha(t) = \infty \quad and \quad \lim_{t \to -\infty} E_\alpha(t) = 0.$$

Consequently, $E_\alpha(\mathbb{R}) = \mathbb{R}_{>0}$ and, due to continuity and monotonicity of E_α, the inverse function of $E_\alpha : \mathbb{R} \to \mathbb{R}_{>0}$, denoted by $\log_\alpha^M : \mathbb{R}_{>0} \to \mathbb{R}$, exists.

Lemma 1.3 [6, Lemma 4] *For any $\alpha \in (0, 1)$, the function $\log_\alpha^M : \mathbb{R}_{>0} \to \mathbb{R}$ satisfies the properties:*

(i) *The function $\log_\alpha^M$ is continuous and strictly monotonically increasing.*
(ii) *For any positive number $\lambda > 0$,*

$$\limsup_{t \to \infty} \frac{1}{t^\alpha} \log_\alpha^M (e^{\lambda^{1/\alpha} t}) = \lambda.$$

(iii) *For any negative number $\lambda < 0$,*

$$\limsup_{t \to \infty} \frac{1}{t^\alpha} \log_\alpha^M \left(\frac{1}{-\lambda \Gamma(1 - \alpha)t^\alpha} \right) = \lambda.$$

The following results are from Kilbas et al. [2].

Proposition 1.1 [2, Proposition 3.23] *The Mittag-Leffler function of a negative argument $E_\alpha(-t)$ is strictly monotonic for all $0 \leq \alpha \leq 1$.*

Lemma 1.4 [2, Lemma 4.25] *For any $\alpha \geq 0$,*

$$E_{\alpha,\alpha}(-t) = -\alpha \frac{d}{dt} E_\alpha(-t), \ \forall t \geq 0.$$

Lemma 1.5 [2, Lemma 4.26] *Let $\beta > \alpha > 0$. Then for any $\alpha \geq 0$,*

$$E_{\alpha,\beta}(-t) = \frac{1}{\alpha \Gamma(\beta - \alpha)} \int_0^1 (1 - s^{1/\alpha})^{\beta-\alpha-1} E_{\alpha,\alpha}(-ts) \, ds, \ \forall t \geq 0.$$

Remark 1.1 It follows from Proposition 1.1, Lemmas 1.4 and 1.5 that $E_{\alpha,\beta}(-t)$ is strictly monotonic on $[0, \infty)$ for all $0 \le \alpha \le 1$ and $\beta > \alpha$.

An important property of the Mittag-Leffler function $E_\alpha(-\gamma t^\alpha)$ with $\gamma > 0$ and $t \ge 0$ is its asymptotic rate of decay.

Lemma 1.6 [1, Theorem 7.3] *Let $\alpha \in (0, 1)$ and $\gamma > 0$. Then $E_\alpha(-\gamma t^\alpha)$ is monotonically decreasing on $[0, \infty)$ and*

$$E_\alpha(-\gamma t^\alpha) = \frac{t^{-\alpha}}{\gamma \, \Gamma(1 - \alpha)} \, (1 + o(1)) \,.$$

In particular, $E_\alpha(-\gamma t^\alpha)$ has no zeros on $[0, \infty)$.

The following identity and its consequences are very important.

Theorem 1.3 [1, Theorem 4.2]

$$E_{\alpha,\beta}(t) = t E_{\alpha,\alpha+\beta}(t) + \frac{1}{\Gamma(\beta)} \tag{1.9}$$

for $\alpha, \beta > 0$ and $t \in \mathbb{R}$.

The proof is by a direct manipulation of the defining series. This is shown here in the following special case, which plays an important role in the sequel.

Lemma 1.7 *Let $\alpha > 0$, $\lambda \in \mathbb{R} \setminus \{0\}$, then*

$$\int_0^t u^{\alpha-1} E_{\alpha,\alpha}(\lambda u^\alpha)\,du = t^\alpha E_{\alpha,\alpha+1}(\lambda t^\alpha) = \frac{1}{\lambda}(E_\alpha(\lambda t^\alpha) - 1), \quad \forall t > 0. \tag{1.10}$$

Proof For any $t > 0$,

$$\begin{aligned}
I(t) &:= \int_0^t u^{\alpha-1} E_{\alpha,\alpha}(\lambda u^\alpha)\,du = \int_0^t u^{\alpha-1} \sum_{k=0}^\infty \frac{(\lambda u^\alpha)^k}{\Gamma(k\alpha + \alpha)}\,du \\
&= \sum_{k=0}^\infty \frac{\lambda^k}{\Gamma(k\alpha + \alpha)} \int_0^t u^{\alpha k + \alpha - 1}\,du \\
&= \sum_{k=0}^\infty \frac{\lambda^k t^{\alpha k + \alpha}}{\Gamma(\alpha k + \alpha)(\alpha k + \alpha)} \\
&= \sum_{k=0}^\infty \frac{\lambda^k t^{\alpha k + \alpha}}{\Gamma(\alpha k + \alpha + 1)}. \tag{1.11}
\end{aligned}$$

From (1.11), it follows that

$$I(t) = t^\alpha \sum_{k=0}^{\infty} \frac{(\lambda t^\alpha)^k}{\Gamma(\alpha k + \alpha + 1)} = t^\alpha E_{\alpha,\alpha+1}(\lambda t^\alpha). \tag{1.12}$$

On the other hand,

$$I(t) = \frac{1}{\lambda} \sum_{k=0}^{\infty} \frac{(\lambda t^\alpha)^{k+1}}{\Gamma(\alpha(k+1)+1)} = \frac{1}{\lambda} \sum_{k=1}^{\infty} \frac{(\lambda t^\alpha)^k}{\Gamma(\alpha k + 1)} = \frac{1}{\lambda}(E_\alpha(\lambda t^\alpha) - 1).$$

By combining (1.11) and (1.12), the proof is complete. $\qquad\square$

It can be shown by a direct calculation that for $\alpha \in (0, 1)$, the function $x(t) = E_\alpha(\gamma t^\alpha)$, $t \geq 0$, is a solution of the linear Caputo FDE

$$\begin{cases} {}^C D_{0+}^\alpha x(t) & = \gamma x(t), \ t > 0, \\ x(0) & = 1. \end{cases}$$

It will be seen in Chap. 3 that this solution is unique. The integral representation of the solution is

$$E_\alpha(\gamma t^\alpha) = 1 + \frac{\gamma}{\Gamma(\alpha)} \int_0^t (t-s)^{\alpha-1} E_\alpha(\gamma s^\alpha)\, ds. \tag{1.13}$$

This gives the following often used inequality:

$$\frac{\gamma}{\Gamma(\alpha)} \int_0^t (t-s)^{\alpha-1} E_\alpha(\gamma s^\alpha)\, ds \leq E_\alpha(\gamma t^\alpha). \tag{1.14}$$

More generally, see [7, Lemma 3.1] and [2, Example 4.9].

Lemma 1.8 *The initial value problem of the scalar linear Caputo FDE of the order $\alpha \in (0, 1)$*

$$\begin{cases} {}^C D_{0+}^\alpha x(t) = \gamma x(t) + h(t), \ t > 0, \\ x(0) = x_0 \end{cases}$$

with $x_0 \in \mathbb{R}$ has the solution

$$x(t) = E_\alpha(\gamma t^\alpha)x_0 + \frac{1}{\Gamma(\alpha)} \int_0^t \frac{1}{(t-s)^{1-\alpha}} E_{\alpha,\alpha}(\gamma(t-s)^\alpha)h(s)\, ds, \ t \geq 0. \tag{1.15}$$

This can be shown using Laplace transforms, e.g., [1, Chap. 7].

1.3.1 Complex-Valued Mittag-Leffler Functions

The domain of definition of Mittag-Leffler functions can be extended to the complex numbers $z \in \mathbb{C}$, in which case $E_\alpha(z)$ takes values in $\mathbb{C}$. Note that its complex-conjugate $\overline{E_\alpha(z)} = E_\alpha(\bar{z})$ for all $z \in \mathbb{C}$.

In particular, $E_\alpha(z)$ has complex-valued (but not real-valued) zeros, see Gorenflo et al. [8, Corollary 3.10, p. 30], which occur in complex-conjugate pairs.

The following property of Mittag-Leffler functions of a complex variable plays an important role in the stability analysis of linear Caputo FDEs in Chap. 5.

Theorem 1.4 [1, Theorem 4.4] *Let $\alpha \in (0, 1]$ and let $z = re^{\iota\phi} \in \mathbb{C}$, where $\iota = \sqrt{-1}$. Then*

(i) $E_\alpha(re^{\iota\phi}) \to 0$ *as* $r \to \infty$ *if* $|\phi| > \alpha\pi/2$;

(ii) $E_\alpha(re^{\iota\phi})$ *remains bounded as* $r \to \infty$ *if* $|\phi| = \alpha\pi/2$;

(iii) $E_\alpha(re^{\iota\phi}) \to \infty$ *as* $r \to \infty$ *if* $|\phi| < \alpha\pi/2$.

1.3.2 Matrix-Valued Mittag-Leffler Functions

Mittag-Leffler functions can be also defined analogously for a matrix $A \in \mathbb{R}^{d \times d}$ and parameters $\alpha > 0$, $\beta > 0$, i.e.,

$$E_{\alpha,\beta}(A) := \sum_{k=0}^{\infty} \frac{A^k}{\Gamma(\alpha k + \beta)}, \qquad E_\alpha(A) := E_{\alpha,1}(A),$$

see, e.g., [1]. Then, specifically,

$$E_\alpha(At) := \sum_{k=0}^{\infty} \frac{A^k t^k}{\Gamma(\alpha k + 1)}, \quad E_{\alpha,\alpha}(At) := \sum_{k=0}^{\infty} \frac{A^k t^k}{\Gamma(\alpha k + \alpha)}.$$

By the same arguments as in the scalar case, the solution of the the linear vector-valued Caputo FDE

$$^C D_{0+}^\alpha x(t) = Ax(t) + h(t), \ t \in (0, T]$$

with the initial condition $x(0) = x_0 \in \mathbb{R}^d$, where $A \in \mathbb{R}^{d \times d}$ and $h := [0, T] \to \mathbb{R}^d$ continuous, is

$$x(t) = E_\alpha(At^\alpha)x_0 + \frac{1}{\Gamma(\alpha)} \int_0^t \frac{1}{(t-s)^{1-\alpha}} E_{\alpha,\alpha}(A(t-s)^\alpha)h(s)\, ds, \ t \in [0, T].$$

$$(1.16)$$

1.4 Endnotes

The monographs of Diethelm [1], Kilbas et al. [2] and Podlubny [5] provide extensive background material and references on fractional calculus and fractional differential equations. They also consider the case where the fractional order α is larger than 1 or some other integer larger than 1.

Gorenflo et al. [8] focuses on Mittag-Leffler functions.

References

1. Diethelm, K.: The analysis of fractional differential equations, Springer Lecture Notes in Mathematics, vol. 2004. Springer, Heidelberg (2010)
2. Kilbas, A.A., Srivastava, H.M., Trujillo, J.J.: Theory and applications of fractional differential equations. North-Holland Mathematics Studies 204, Elsevier Science B.V., Amsterdam (2006) MR2218073
3. Miller, R.K., Sell, G.R.: Volterra integral equations and topological dynamics. Memoir Amer. Math. Soc. **102** (1970)
4. Vainikko, G.: Which functions are fractionally differentiable? J. Anal. Applns **35**, 465–487 (2016)
5. Podlubny, I.: Fractional differential equations. An introduction to fractional derivatives, fractional differential equations, to methods of their solution and some of their applications. Academic Press, Inc., San Diego, CA (1999)
6. Cong, N.D., Doan, T.S., Tuan, H.T.: On fractional Lyapunov exponent for solutions of linear fractional differential equations. Fract. Calculus Appl. Anal. **17**(2), 285–306 (2014)
7. Cong, N.D., Tuan, H.T.: Generation of nonlocal dynamical systems by fractional differential equations. J. Integral Equ. Appl. **29**, 585–608 (2017). https://doi.org/10.1216/JIE-2017-29-4-585
8. Gorenflo, R., Kilbas, A.A., Mainardi, F., Rogosin, S.V.: Mittag-Leffler functions. Related topics and applications. Springer-Verlag, Berlin Heidelberg (2014)

Chapter 2
Useful Inequalities

Abstract Some useful inequalities in the qualitative analysis of fractional differential equations, including the Tuan–Trinh inequality and a generalized Gronwall inequality are presented, as well as some fundamental comparison principles in the fractional framework.

Keywords Tuan and Trinh inequality · Generalized Gronwall inequality · Comparison arguments

The fractional derivative does not have the convenient properties and geometric association that the ordinary derivative has, e.g., a negative fractional derivative does not necessarily imply that the function is decreasing. In particular, the usual rules of differential calculus do not always apply.

Nevertheless, for $\alpha \in (0, 1)$, it follows from Tuan and Trinh [1, Theorem 2] that solutions of the Caputo FDE

$$^{C}D_{0+}^{\alpha} x(t) = g(x(t)), t > 0, \tag{2.1}$$

satisfy

$$^{C}D_{0+}^{\alpha} \|x(t)\|^{2} \leq 2\langle x(t), {}^{C}D_{0+}^{\alpha} x(t) \rangle. \tag{2.2}$$

An earlier proof by Aguila-Camacho et al. [2, Lemma 1] requires the paths to be continuously differentiable, so is not valid for the solutions of Caputo FDEs.

Based on the inequality (2.2), we have

$$^{C}D_{0+}^{\alpha} \|x(t)\|^{2} \leq 2\langle x(t), g(x(t)) \rangle$$

along the solutions of (2.1), so the Caputo differential inequality

$$^{C}D_{0+}^{\alpha} \|x(t)\|^{2} \leq 2a - 2b\|x(t)\|^{2}, \tag{2.3}$$

T. S. Doan et al., *Attractors of Caputo Fractional Differential Equations*,
SpringerBriefs in Mathematics, https://doi.org/10.1007/978-3-032-05511-8_2

holds, when the vector field g satisfies the dissipativity condition

$$\langle x, g(x) \rangle \leq a - b\|x\|^2. \tag{2.4}$$

The inequality (2.3) has many uses in investigating qualitative and dynamical properties of the solutions of dissipative Caputo FDEs.

Throughout this chapter (and book) it is always assumed that the fractional order $\alpha \in (0, 1)$.

2.1 An Inequality of Tuan and Trinh

Tuan and Trinh [1, Theorem 2] actually proved a more general inequality than (2.2). Their result provides an upper bound of the fractional derivative of a composite function.

Recall that $I^\alpha_{0+} C([0, T], \mathbb{R}^d)$ denotes the space of functions $\varphi : [0, T] \to \mathbb{R}^d$ such that there exists a function $\psi \in C([0, T], \mathbb{R}^d)$ satisfying $\varphi = I^\alpha_{0+} \psi$.

Theorem 2.1 *For a given $x_0 \in \mathbb{R}^d$, let $u \in \{x_0\} + I^\alpha_{0+} C([0, T], \mathbb{R}^d)$ and suppose that $V : \mathbb{R}^d \to \mathbb{R}$ satisfies the following conditions:*

(V.1) the function V is convex on $\mathbb{R}^d$ and $V(0) = 0$;

(V.2) the function V is differentiable on $\mathbb{R}^d$.

Then

$$^C D^\alpha_{0+} V(u(t)) \leq \langle \nabla V(u(t)), ^C D^\alpha_{0+} u(t) \rangle \tag{2.5}$$

for all $t \in [0, T]$, where ∇V is the gradient of the function V.

Proof Since $u \in \{x_0\} + I^\alpha_{0+} C([0, T], \mathbb{R}^d)$, there exists a function $\psi \in C([0, T], \mathbb{R}^d)$ such that $u = x_0 + I^\alpha_{0+} \psi$.

By [3, Proposition 6.4, pp. 479], the Caputo fractional derivative $^C D^\alpha_{0+} u$ exists and is continuous on $[0, T]$. Moreover, by Theorem 1.1, this derivative has the representation

$$^C D^\alpha_{0+} u(0) := \Gamma(\alpha + 1)\gamma, \tag{2.6}$$

and

$$^C D^\alpha_{0+} u(t) = \frac{u(t) - u(0)}{\Gamma(1 - \alpha)t^\alpha} + \frac{\alpha}{\Gamma(1 - \alpha)} \int_0^t \frac{u(t) - u(\tau)}{(t - \tau)^{\alpha+1}} d\tau \tag{2.7}$$

for $0 < t \leq T$.

It then follows from (V.1), (V.2) and direct computation that

$$\lim_{t \to 0} \frac{V(u(t)) - V(u(0))}{t^\alpha} = \langle \nabla V(u(0)), \gamma \rangle.$$

Now

$$\sup_{0 < t \le T} \left\| \int_{\theta t}^{t} (t - \tau)^{-\alpha-1} (u(t) - u(\tau)) d\tau \right\| \to 0 \quad \text{as} \quad \theta \to 1,$$

implies that the limit

$$\sup_{0 < t \le T} \left| \int_{\theta t}^{t} (t - \tau)^{-\alpha-1} (V(u(t)) - V(u(\tau))) d\tau \right| \to 0$$

as $\theta \to 1$ holds. Together with Theorem 1.1 this gives

$$^{C}D_{0+}^{\alpha} V(u(0)) = \Gamma(\alpha + 1) \langle \nabla V(u(0)), \gamma \rangle \tag{2.8}$$

and

$$^{C}D_{0+}^{\alpha} V(u(t)) = \frac{V(u(t)) - V(u(0))}{\Gamma(1 - \alpha)t^\alpha} + \frac{\alpha}{\Gamma(1 - \alpha)} \int_{0}^{t} \frac{V(u(t)) - V(u(\tau))}{(t - \tau)^{\alpha+1}} d\tau \tag{2.9}$$

for $0 < t \le T$.

Moreover, from (2.6) and (2.8)

$$^{C}D_{0+}^{\alpha} V(u(0)) = \langle \nabla V(u(0)), {}^{C}D_{0+}^{\alpha} u(0) \rangle. \tag{2.10}$$

Then, using the representations (2.7) and (2.9) leads to

$$^{C}D_{0+}^{\alpha} V(u(t)) - \langle \nabla V(u(t)), {}^{C}D_{0+}^{\alpha} u(t) \rangle$$

$$= \frac{V(u(t)) - V(u(0)) - \langle \nabla V(u(t)), u(t) - u(0) \rangle}{\Gamma(1 - \alpha)t^\alpha}$$

$$+ \frac{\alpha}{\Gamma(1 - \alpha)} \int_{0}^{t} \frac{V(u(t)) - V(u(\tau))}{(t - \tau)^{\alpha+1}} d\tau$$

$$- \frac{\alpha}{\Gamma(1 - \alpha)} \int_{0}^{t} \frac{\langle \nabla V(u(t)), u(t) - u(\tau) \rangle}{(t - \tau)^{\alpha+1}} d\tau \tag{2.11}$$

for $0 < t \le T$.

Since V is convex and differentiable, it follows by [4, Theorem 25.1, pp. 242] that

$$V(u(t)) - V(u(\tau)) - \langle \nabla V(u(t)), u(t) - u(\tau) \rangle \le 0$$

for all $0 \leq \tau \leq t \leq T$. Together with (2.10) and (2.11) this implies that

$$^C D_{0+}^\alpha V(u(t)) \leq \langle \nabla V(u(t)), ^C D_{0+}^\alpha u(t)\rangle, \quad \forall t \in [0, T].$$

This completes the proof of Theorem 2.1. $\square$

Remark 2.1 In general, the solutions to fractional differential equations are not always differentiable. For example, consider the equation

$$^C D_{0+}^\alpha x(t) = f(t), \quad t > 0, \quad x(0) = 0, \tag{2.12}$$

where

$$f(t) = \begin{cases} t^\beta, & t \in [0, 1], 0 < \beta < 1 - \alpha, \\ 1, & t \geq 1. \end{cases}$$

It is clear that the solution $\varphi(\cdot, 0)$ to (2.12) is continuous on $\mathbb{R}_{\geq 0}$, but not differentiable at $t = 0$.

2.2   Some Caputo Differential Inequalities

Lemma 2.1 *Let* $w(\cdot), {}^C D_{0+}^\alpha w(\cdot) \in C([0, T]; \mathbb{R})$ *and assume that*

$$^C D_{0+}^\alpha w(t) \leq \gamma w(t), \forall t \in (0, T].$$

Then,
$$w(t) \leq w(0) E_\alpha(\gamma t^\alpha), \forall t \in [0, T]. \tag{2.13}$$

Proof There exists $m(\cdot) \in C([0, T]; \mathbb{R}_{\geq 0})$ such that

$$^C D_{0+}^\alpha w(t) = \gamma w(t) - m(t), \forall t \in (0, T].$$

From Lemma 1.8, the solution of this linear Caputo FDE is

$$w(t) = w(0) E_\alpha(\gamma t^\alpha) - \int_0^t \frac{m(s)}{(t - s)^{1-\alpha}} E_{\alpha,\alpha}(\gamma(t - s)^\alpha) ds.$$

It follows from Lemmas 1.1, 1.4 and 1.5 that $E_{\alpha,\alpha}(\gamma(t - s)^\alpha) \geq 0$ for all $s \geq 0$ and $t \geq s$, so

$$w(t) \leq w(0) E_\alpha(\gamma t^\alpha), \forall t \in [0, T]. \tag{2.14}$$

This completes the proof of Lemma 2.1. $\square$

Corollary 2.1 *Let $w(\cdot) \in C([0, T]; \mathbb{R})$ and*

$$^{C}D_{0+}^{\alpha} w(t) \leq 0, \forall t \in (0, T].$$

Then,

$$w(t) \leq w(0), \forall t \in [0, T].$$

Proof The proof is obtained by applying Lemma 2.1 for $\gamma = 0$.

By using the similar arguments as in the proof of Lemma 2.1 and Corollary 2.1, it is easy to derive the following results.

Lemma 2.2 *Let $w(\cdot), {}^{C}D_{0+}^{\alpha} w(\cdot) \in C([0, T]; \mathbb{R})$ and assume that*

$$^{C}D_{0+}^{\alpha} w(t) \geq \gamma w(t), \forall t \in (0, T].$$

Then,

$$w(t) \geq w(0) E_{\alpha}(\gamma t^{\alpha}), \forall t \in [0, T]. \tag{2.15}$$

Corollary 2.2 *Let $w(\cdot) \in C([0, T]; \mathbb{R})$ and*

$$^{C}D_{0+}^{\alpha} w(t) \geq 0, \forall t \in (0, T].$$

Then

$$w(t) \geq w(0), \forall t \in [0, T].$$

The following elementary inequality for Caputo fractional differential inequalities of order $\alpha \in (0, 1)$ has appeared in one form or another in the literature, but does not seem to be widely known. The following version is from [5, Lemma 1].

Lemma 2.3 *Let $a \geq 0$, $b > 0$. Suppose that $w(\cdot), {}^{C}D_{0+}^{\alpha} w(\cdot) \in C([0, T]; \mathbb{R})$ and*

$$^{C}D_{0+}^{\alpha} w(t) \leq 2a - 2bw(t), \forall t \in (0, T]. \tag{2.16}$$

Then

$$w(t) \leq w(0) E_{\alpha}(-2bt^{\alpha}) + \frac{a}{b}\left(1 - E_{\alpha}(-2bt^{\alpha})\right), \forall t \in [0, T]. \tag{2.17}$$

Proof By the assumption that $w(\cdot), {}^{C}D_{0+}^{\alpha} w(\cdot) \in C([0, T]; \mathbb{R})$ and the inequality (2.16), we can find a function $m(\cdot) \in C([0, T]; \mathbb{R}_{\geq 0})$ such that for all $t \in (0, T]$:

$$^{C}D_{0+}^{\alpha} w(t) + 2bw(t) = 2a - m(t).$$

It deduces from Lemma 1.8 that the solution of the linear Caputo FDE above is

$$w(t) = w_0 E_\alpha(-2bt^\alpha) + 2a \int_0^t \frac{1}{(t-s)^{1-\alpha}} E_{\alpha,\alpha}(-2b(t-s)^\alpha) ds$$

$$- \int_0^t \frac{m(s)}{(t-s)^{1-\alpha}} E_{\alpha,\alpha}(-2b(t-s)^\alpha) ds, \forall t \in [0,T],$$

which together with the fact that $E_{\alpha,\alpha}(-2b(t-s)^\alpha) \geq 0$ implies

$$w(t) \leq w(0) E_\alpha(-2bt^\alpha) + 2a \int_0^t \frac{1}{(t-s)^{1-\alpha}} E_{\alpha,\alpha}(-2b(t-s)^\alpha) ds, \forall t \in [0,T].$$

$$(2.18)$$

On the other hand, from the integral representation of the Mittag-Leffler function (see Lemma 1.7), it follows that

$$g(t) := \int_0^t \frac{1}{(t-s)^{1-\alpha}} E_{\alpha,\alpha}(-2b(t-s)^\alpha) ds = t^\alpha E_{\alpha,\alpha+1}(-2bt^\alpha). \qquad (2.19)$$

Thus, by (2.18) and (2.19),

$$w(t) \leq w_0 E_\alpha(-2bt^\alpha) + 2at^\alpha E_{\alpha,\alpha+1}(-2bt^\alpha).$$

Furthermore, by Theorem 1.3,

$$E_\alpha(\tau) = E_{\alpha,1}(\tau) = \tau E_{\alpha,\alpha+1}(\tau) + \frac{1}{\Gamma(1)} = \tau E_{\alpha,\alpha+1}(\tau) + 1, \forall \tau \in \mathbb{R}.$$

So with $\tau = -2bt^\alpha$, then

$$2bt^\alpha E_{\alpha,\alpha+1}(-2bt^\alpha) = 1 - E_\alpha(-2bt^\alpha).$$

Hence, for $t \geq 0$

$$w(t) \leq w(0) E_\alpha(-2bt^\alpha) + \frac{a}{b}\left(1 - E_\alpha(-2bt^\alpha)\right).$$

$$\square$$

Remark 2.2 The proof of inequality (2.17) is partly based on the proof of Wang & Xiao [6, Theorem 1], although the inequality was not derived there. In their proof Wang and Xiao showed that the function $g(t)$ defined in (2.19) is positive and monotonically increasing to the value $\frac{1}{2b}$. In particular, they showed that its derivative satisfies

$$g'(t) = \frac{d}{dt}\left[t^\alpha E_{\alpha,\alpha+1}(-2b(t-s)^\alpha)\right] = t^{\alpha-1} E_{\alpha,\alpha}(-2bt^\alpha) > 0$$

and used an asymptotic expansion for the Mittag-Leffler function [7, 8] to obtain

$$\lim_{t \to \infty} g(t) = \lim_{t \to \infty} t^\alpha \left(\frac{1}{2bt^\alpha} + O\left(\frac{1}{t^{2\alpha}} \right) \right) = \frac{1}{2b}.$$

A variation of inequality (2.17) was established by Alikhanov [9, Lemma 2] for time dependent coefficients. Neither Alikhanov [9] nor Wang & Xiao [6] established the actual inequality (2.17).

2.3 The Fractional Gronwall Inequality

The following fractional version of Gronwall's inequality is from Diethelm [10, Lemma 6.19].

Lemma 2.4 *Let $\alpha \in (0, 1)$, μ, ν, $T \in \mathbb{R}_{\geq 0}$, and let $x : [0, T] \to \mathbb{R}_{\geq 0}$ be a continuous function satisfying the inequality*

$$x(t) \leq \mu + \frac{\nu}{\Gamma(\alpha)} \int_0^t (t - s)^{\alpha-1} x(s)\, ds, \forall t \in [0, T].$$

Then

$$x(t) \leq \mu E_\alpha(\nu t^\alpha), \forall t \in [0, T]. \tag{2.20}$$

The following generalisation of the fractional Gronwall inequality (2.20) will be needed in Sect. 3.7, where the function τ is a step function.

Lemma 2.5 *Let $\alpha \in (0, 1)$ and T, K, $L > 0$ be positive constants. Suppose that $h :$ $[0, T] \to \mathbb{R}_{\geq 0}$ is a continuous function satisfying that*

$$h(t) \leq K + \frac{L}{\Gamma(\alpha)} \int_0^t (t - s)^{\alpha-1} h(\tau(s))ds, \forall t \in [0, T], \tag{2.21}$$

where $0 \leq \tau(s) \leq s$ for all $s \in [0, T]$. Then,

$$h(t) \leq K E_\alpha(Lt^\alpha), \forall t \in [0, T].$$

Proof Choose and fix $\varepsilon > 0$. Let $\varphi(t) := (K + \varepsilon)E_\alpha(Lt^\alpha)$, $t \in [0, T]$. Then $\varphi(t)$ is the unique solution of the following initial value problem on $[0, T]$

$$^C D_{0+}^\alpha x(t) = Lx(t), \qquad x(0) = K + \varepsilon.$$

Thus, $\varphi(t)$ also satisfies the integral equation

$$\varphi(t) = K + \varepsilon + \frac{L}{\Gamma(\alpha)} \int_0^t (t - s)^{\alpha-1} \varphi(s)ds, \quad \forall t \in [0, T].$$

Assume that there exists $t \in (0, T]$ such that $h(t) \geq \varphi(t)$. Let

$$t_0 := \inf\{t \in (0, T] : h(t) \geq \varphi(t)\}. \tag{2.22}$$

Since $h(0) \leq K < \varphi(0)$ and $h(t), \varphi(t)$ are continuous functions on $[0, T]$ it follows that

$$h(t_0) = \varphi(t_0), \quad h(t) < \varphi(t) \quad \text{for all } t \in [0, t_0). \tag{2.23}$$

This together with (2.21) and the fact that the function $E_\alpha(\cdot)$ is monotone increasing on $\mathbb{R}_{\geq 0}$ imply that

$$
\begin{aligned}
h(t_0) &\leq K + \frac{L}{\Gamma(\alpha)} \int_0^{t_0} (t_0 - s)^{\alpha-1} h(\tau(s)) \, ds \\
&< K + \frac{L}{\Gamma(\alpha)} \int_0^{t_0} (t_0 - s)^{\alpha-1} (K + \varepsilon) E_\alpha(L(\tau(s))^\alpha) \, ds \\
&\leq K + \frac{L}{\Gamma(\alpha)} \int_0^{t_0} (t_0 - s)^{\alpha-1} (K + \varepsilon) E_\alpha(Ls^\alpha) \, ds \\
&= K + \varepsilon + \frac{L}{\Gamma(\alpha)} \int_0^{t_0} (t_0 - s)^{\alpha-1} \varphi(s) \, ds \\
&= \varphi(t_0),
\end{aligned}
$$

which contradicts (2.23). Therefore,

$$h(t) < (K + \varepsilon) E_\alpha(Lt^\alpha), \ \forall t \in [0, T].$$

Letting $\varepsilon \to 0$, gives the desired estimate. $\qquad\square$

2.4   Comparision Principles

Consider the Caputo FDE (2.1), i.e.,

$$^{C}D_{0+}^{\alpha} x(t) = g(x(t)), \ \forall t \in (0, T].$$

Lower and upper solutions of this equation have an important role in many proofs.

Definition 2.1 Functions $v, w \in C([0, T], \mathbb{R})$ satisfying

(i) $v(t) \leq x_0 + \dfrac{1}{\Gamma(\alpha)} \int_0^t (t - s)^{\alpha-1} g(v(s)) ds$ for all $t \in [0, T]$,

(ii) $w(t) \geq x_0 + \dfrac{1}{\Gamma(\alpha)} \int_0^t (t - s)^{\alpha-1} g(w(s)) ds$ for all $t \in [0, T]$

are called, respectively, lower and upper solutions of the Caputo FDE (2.1).

The following lemma is a small modification of [11, Theorem 2.1]. The proof is given here for convenience of the reader.

Lemma 2.6 (Comparision Principle) *Let $0 < \alpha < 1$ and let $v_1, w_1 \in C([0, T], \mathbb{R})$ satisfy*

$$v_1(t) < w_1(t) \quad \text{for all } t \in [0, T]. \tag{2.24}$$

In addition, suppose that $g \in C(\mathbb{R}, \mathbb{R})$ is nondecreasing and that $v, w \in C([0, T], \mathbb{R})$ are such that

(i) $v(t) \leq v_1(t) + \dfrac{1}{\Gamma(\alpha)} \int_0^t (t - s)^{\alpha-1} g(v(s))ds, \; \forall \in [0, T],$

(ii) $w(t) \geq w_1(t) + \dfrac{1}{\Gamma(\alpha)} \int_0^t (t - s)^{\alpha-1} g(w(s))ds, \; \forall t \in [0, T].$

Then,

$$v(t) < w(t) \quad \text{for all } t \in [0, T]. \tag{2.25}$$

Proof Suppose that (2.25) is not true. Then, from the continuity of $v(\cdot)$ and $w(\cdot)$ there exists $0 < t_1 \in [0, T]$ such that

$$v(t_1) = w(t_1), \quad \text{and} \quad v(t) < w(t) \text{ for all } 0 \leq t < t_1. \tag{2.26}$$

Since g is nondecreasing nature, (2.26) and (2.24) give

$$\begin{aligned}
v(t_1) &\leq v_1(t_1) + \frac{1}{\Gamma(\alpha)} \int_0^{t_1} (t_1 - s)^{\alpha-1} g(v(s))ds \\
&\leq v_1(t_1) + \frac{1}{\Gamma(\alpha)} \int_0^{t_1} (t_1 - s)^{\alpha-1} g(w(s))ds \\
&< w_1(t_1) + \frac{1}{\Gamma(\alpha)} \int_0^{t_1} (t_1 - s)^{\alpha-1} g(w(s))ds \\
&\leq w(t_1).
\end{aligned}$$

This contradicts the assumption that $v(t_1) = w(t_1)$ in (2.26). Hence the assertion (2.25) is valid. $\qquad\square$

The proof of the next lemma is obtained by using arguments as in the proof of [12, Lemma 2.1].

Lemma 2.7 *Let $m : [0, T] \to \mathbb{R}$ be continuous and let its Caputo derivative ${}^C D_{0+}^\alpha m$ exist on the interval $(0, T]$. If there is a $t_0 \in (0, T]$ such that*

$$m(t) \leq 0, \; \forall t \in [0, t_0), \; \text{and } m(t_0) = 0,$$

then ${}^C D_{0+}^\alpha m(t_0) \geq 0$.

Lemma 2.8 *Let $L : \mathbb{R} \to \mathbb{R}$ be continuous and nonincreasing and let $m_1 : [0, T] \to \mathbb{R}$, $m_2 : [0, T] \to \mathbb{R}$ be continuous. Assume that $^C D_{0+}^\alpha m_1$, $^C D_{0+}^\alpha m_2$ exist on $(0, T]$. If*

$$^C D_{0+}^\alpha m_1(t) \geq L(m_1(t)), \quad t \in (0, T], \quad m_1(0) \geq m_0, \tag{2.27}$$

$$^C D_{0+}^\alpha m_2(t) \leq L(m_2(t)), \quad t \in (0, T], \quad m_2(0) \leq m_0, \tag{2.28}$$

then $m_1(t) \geq m_2(t)$ for all $t \in (0, T]$.

Proof Assume first that one of the inequalities in (2.27) and (2.28) is strict, say $^C D_{0+}^\alpha m_2(t) < L(m_2(t))$ and $m_2(0) < m_0 \leq m_1(0)$. Then, for all $t \in [0, T]$, the following inequality holds

$$m_2(t) < m_1(t).$$

Indeed, suppose that there is a $t_0 \in (0, T]$ such that $m_2(t_0) = m_1(t_0)$ and $m_2(t) < m_1(t)$ on the interval $[0, t_0)$. Set $m(t) = m_2(t) - m_1(t)$ it follows that $m(t_0) = 0$ and $m(t) < 0$ for $t \in [0, t_0)$. Then Lemma 2.7 implies that $^C D_{0+}^\alpha m(t_0) \geq 0$. However, $m_2(t_0) = m_1(t_0)$, so

$$L(m_2(t_0)) > {}^C D_{0+}^\alpha m_2(t_0) \geq {}^C D_{0+}^\alpha m_1(t_0) \geq L(m_1(t_0)) = L(m_2(t_0)),$$

a contradiction. Hence, $m_2(t) < m_1(t)$ on $[0, T]$.

Now assume that the inequalities in (2.27) are non-strict. It will be shown that $m_2(t) \leq m_1(t)$ for all $t \in [0, T]$. Set $m_1^\varepsilon(t) = m_1(t) + \varepsilon\lambda(t)$ where $\varepsilon > 0$ and $\lambda(t) = E_\alpha(t^\alpha)$. Since $\lambda(\cdot)$ is positive and $L(\cdot)$ is non-increasing,

$$
\begin{aligned}
^C D_{0+}^\alpha m_1^\varepsilon(t) &= {}^C D_{0+}^\alpha m_1(t) + \varepsilon\lambda(t) \\
&\geq L(m_1(t)) + \varepsilon\lambda(t) \\
&= L(m_1^\varepsilon(t)) + L(m_1(t)) - L(m_1^\varepsilon(t)) + \varepsilon\lambda(t) \\
&\geq L(m_1^\varepsilon(t)) + \varepsilon\lambda(t) \\
&> L(m_1^\varepsilon(t)), \quad \forall t \in (0, T].
\end{aligned}
\tag{2.29}
$$

Due to (2.29) and the result above for strict inequalities, it follows that $m_2(t) < m_1^\varepsilon(t)$ for all $t \in [0, T]$. Consequently, letting $\varepsilon \to 0$ leads to $m_2(t) \leq m_1(t)$, $\forall t \in [0, T]$. The proof is complete. $\qquad\square$

2.5   Endnotes

The inequality of Tuan and Trinh [1, Theorem 2] plays a critical role in the analysis of dissipative Caputo FDEs. An immediate consequence is the inequality in Lemma 2.3, which is taken from Kloeden [5, Lemma 1].

Variations of the Gronwall inequality, both continuous and discrete time, have been established, see, e.g., Dixon [13], Doan et al. [14], Ye et al. [15]. For generalizations to Henry-Gronwall inequalities for Caputo FDEs see Zhu [16].

The proof of Lemma 2.8 is from Cong et al. [17, Proposition 26]. It is based on that of [12, Theorem 2.3], but does not require continuous differentiability of $m_1(\cdot)$, $m_2(\cdot)$ and the Lipschitz property of $L(\cdot)$ used there.

References

1. Tuan, H.T., Trinh, H.: Stability of fractional-order nonlinear systems by Lyapunov direct method. IET Control Theory Appl. **12**, 2417–2422 (2018)
2. Aguila-Camacho, N., Duar-Mermoud, M.A., Javier, A.: Gallegos, Lyapunov functions for fractional order systems. Commun. Nonlinear Sci. Numer. Simulat **19**, 2951–2957 (2014). https://doi.org/10.1016/j.cnsns.2014.01.022
3. Vainikko, G.: Which functions are fractionally differentiable? J. Anal. Appl. **35**, 465–487 (2016)
4. Rockafellar, R.T.: Convex Analysis. Princeton University Press, Princeton, New Jersey (1972)
5. Kloeden, P.E.: An elementary inequality for dissipative Caputo fractional differential equations. Fract. Calc. Appl. Anal. **26**, 2166–2174 (2023). https://doi.org/10.1007/s13540-023-00192-x
6. Wang, D., Xiao, A.: Dissipativity and contractivity for fractional-order systems. Nonlinear Dyn. **80**(1–2), 287–294 (2015). https://doi.org/10.1007/s11071-014-1868-1
7. Kilbas, A.A., Srivastava, H.M., Trujillo, J.J.: Theory and applications of fractional differential equations. North-Holland Mathematics Studies 204, Elsevier Science B.V., Amsterdam (2006) MR2218073
8. Podlubny, I.: Fractional differential equations. An introduction to fractional derivatives, fractional differential equations, to methods of their solution and some of their applications. Academic Press, Inc., San Diego, CA (1999)
9. Alikhanov, A.A.: A priori estimates for solutions of boundary value problems for fractional-order equations. Differ. Equ. **46**, 660–666 (2010)
10. Diethelm, K.: The analysis of fractional differential equations. Springer Lecture Notes in Mathematics, vol. 2004. Springer, Heidelberg (2010)
11. Lakshmikantham, V., Vatsala, A.S.: Basic theory of fractional differential equations. Nonlinear Anal. **69**, 2677–2682 (2008)
12. Ramirez, J.D., Vatsala, A.S.: Generalized monotone iterative technique for Caputo fractional differential equation with periodic boundary condition via initial value problem. Int. J. Differ. Eqn. (2012) (SI2), 1–17 (2012). https://doi.org/10.1155/2012/842813
13. Dixon, J.: On the order of the error in discretization methods for weakly singular second kind non-smooth solutions. BIT **25**, 623–634 (1986)
14. Doan, T.S., Huong, T.P., Kloeden, P.E.: The theta-scheme for solving Caputo fractional differential equations. Electron. J. Differ. Equ. **2025**(05), 1–13 (2025)
15. Ye, H., Gao, J., Ding, Y.: A generalization Gronwall inequality and its application to a fractional differential equation. J. Math. Anal. Appl. **328**, 1075–1081 (2007)
16. Tao, Z.: New Henry-Gronwall integral inequalities and their applications to fractional differential equations. Bulletin Braz. Math. Soc. New Ser. **49**, 647–657 (2018)
17. Cong, N.D., Tuan, H.T., Hieu, T.: On asymptotic properties of solutions to fractional differential equations. J. Math. Anal. Appl. **484**, 123759 (2020)

Chapter 3
Existence and Uniqueness of Solutions

Abstract The existence and uniqueness of solutions to integral equations with a special degenerate kernel associated with the Caputo fractional integral is established using an approach based on fixed point theorems. Several properties of solutions are also examined, such as Hölder continuity, continuous dependence on the input data, and continuous dependence on the fractional order of the derivative. In addition, the invariance under time shifts in autonomous systems for solutions of fractional-order initial value problems, as well as explicit solution representations in terms of Mittag-Leffler functions (via the variation of constants formula), are considered. Finally, some results on the existence of solutions to boundary value problems for scalar Caputo fractional differential equations are presented.

Keywords Integral equations · Caputo fractional differential equations · Existence and uniqueness of solutions · Initial value problem · Boundary value problems · Holder continuity of solutions · Continuous dependence on the input data · Continuous dependence on the fractional order · Variation of constants formula

A unique global solution exists for the initial value problem for a Caputo fractional differential equation (FDE) of order $\alpha \in (0, 1)$ in $\mathbb{R}^d$

$$^{C}D_{0+}^{\alpha} x(t) = g(x(t)), \ \forall t > 0, \tag{3.1}$$

$$x(0) = x_0, \tag{3.2}$$

when vector field $g : \mathbb{R}^d \to \mathbb{R}^d$ is globally Lipschitz, i.e., satisfies the assumption:

Assumption 3.1 There exists $L > 0$ such that for all $x, y \in \mathbb{R}^d$,

$$\|g(x) - g(y)\| \leq L\|x - y\|. \tag{3.3}$$

For the initial value $x_0 \in \mathbb{R}^d$, the corresponding integral equation is

$$x(t) = x_0 + \frac{1}{\Gamma(\alpha)} \int_0^t (t - s)^{\alpha-1} g(x(s))ds, \ \forall t \geq 0. \tag{3.4}$$

© The Author(s), under exclusive license to Springer Nature Switzerland AG 2025
T. S. Doan et al., *Attractors of Caputo Fractional Differential Equations*,
SpringerBriefs in Mathematics, https://doi.org/10.1007/978-3-032-05511-8_3

The desired existence and uniqueness theorem is:

Theorem 3.1 *Assume that the vector field g satisfies the global Lipschitz condition in Assumption 3.1. Then, for any $T > 0$ and for each initial value $x_0 \in \mathbb{R}^d$, the integral equation (3.4) has a unique solution $\varphi(\cdot, x_0)$ on the interval $[0, T]$. Moreover, the solution $\varphi(\cdot, x_0)$ is continuous in $x_0 \in \mathbb{R}^d$.*

3.1 Global Existence and Uniqueness of Solutions

For latter purposes a more general theorem than Theorem 3.1 will be proved here, namely for the corresponding Volterra integral equation [1]

$$x(t) = f(t) + \frac{1}{\Gamma(\alpha)} \int_0^t (t - s)^{\alpha-1} g(x(s)) ds, \ t \geq 0, \tag{3.5}$$

where $f : \mathbb{R}^+ \to \mathbb{R}^d$ is a continuous function. This reduces to the integral equation representation (3.4) of (2.1) when $f(t) \equiv x_0$.

The following lemma is needed.

Lemma 3.1 *Let $\theta > 0$. The function $r_\theta(t) := (t + \theta)^\alpha - t^\alpha$, $t \geq 0$, is montonically decreasing from the maximum value θ^α. In particular, $0 < r_\theta(t) \leq \theta^\alpha$ for all $t \geq 0$.*

Proof Note that $r_\theta(0) = \theta^\alpha$ and that the derivative $r_\theta'(t) = \frac{1}{\alpha}\left((t + \theta)^{\alpha-1} - t^{\alpha-1}\right) < 0$ with $r_\theta'(t) \to 0^-$ as $t \to \infty$. $\qquad\Box$

Theorem 3.2 *Assume that the vector field g satisfies the global Lipschitz condition in Assumption 3.1. Then for any $T > 0$ and for each $f \in C([0, T], \mathbb{R}^d)$, the integral equation (3.5) has a unique solution $x_f(t)$ on the interval $[0, T]$.*

Moreover, the solution $x_f(t)$ depends continuously on $f \in C([0, T], \mathbb{R}^d)$ in the supremum norm.

The proof of Theorem 3.2 is based on a contraction mapping argument for an operator $\mathcal{T}_f$ defined on $C([0, T], \mathbb{R}^d)$ by

$$\mathcal{T}_f\xi(t) := f(t) + \frac{1}{\Gamma(\alpha)} \int_0^t (t - \tau)^{\alpha-1} g(\xi(\tau)) \, d\tau, \qquad \xi \in C([0, T], \mathbb{R}^d), \quad (3.6)$$

for a fixed $f \in C([0, T], \mathbb{R}^d)$. This gives only local existence if the usual supremum norm on continuous functions is used [2, Theorem 6.2]. Unlike for solutions of ODEs, these local Caputo solutions cannot in general be patched together to provide a global solution for the Caputo FDE. Instead, following [3, Theorem 3.1] or [4, Theorem 1], the Banach space $C([0, T], \mathbb{R}^d)$ will be used with a suitable Bielecki weighted norm

$$\|\xi\|_\gamma := \sup_{t\in[0,T]} \frac{\|\xi(t)\|}{E_\alpha(\gamma t^\alpha)} \quad \text{for all } \xi \in C([0,T],\mathbb{R}^d), \tag{3.7}$$

where $\gamma > 0$ is a suitable constant (to be determined later in the proof) and the weight function $E_\alpha(\cdot)$ is a Mittag-Leffler function.

The proof of Theorem 3.2 is given in the following three lemmas.

Lemma 3.2 *For any $f \in C([0,T],\mathbb{R}^d)$ the operator $\mathcal{T}_f : C([0,T],\mathbb{R}^d) \to C([0,T],\mathbb{R}^d)$ in (3.6) is well-defined.*

Proof Let $f \in C([0,T],\mathbb{R}^d)$ be arbitrary but fixed and $\xi \in C([0,T],\mathbb{R}^d)$. Then, for $t, t+r \in [0,T]$ with $r > 0$,

$$\mathcal{T}_f\xi(t+r) - \mathcal{T}_f\xi(t) = f(t+r) - f(t) + \frac{1}{\Gamma(\alpha)} \int_0^{t+r} (t+r-\tau)^{\alpha-1} g(\xi(\tau))\, d\tau$$

$$- \frac{1}{\Gamma(\alpha)} \int_0^t (t-\tau)^{\alpha-1} g(\xi(\tau))\, d\tau$$

$$= f(t+r) - f(t) + \frac{1}{\Gamma(\alpha)} \int_t^{t+r} (t+r-\tau)^{\alpha-1} g(\xi(\tau))\, d\tau$$

$$+ \frac{1}{\Gamma(\alpha)} \int_0^t \left((t+r-\tau)^{\alpha-1} - (t-\tau)^{\alpha-1} \right) g(\xi(\tau))\, d\tau.$$

Put $M := \max_{t\in[0,T]} \|g(\xi(t))\| < \infty$. Then,

$$\left\| \int_t^{t+r} (t+r-\tau)^{\alpha-1} g(\xi(\tau))\, d\tau \right\| \leq \int_t^{t+r} (t+r-\tau)^{\alpha-1} \|g(\xi(\tau))\|\, d\tau$$

$$\leq M \int_t^{t+r} (t+r-\tau)^{\alpha-1} d\tau \leq \frac{M}{\alpha} r^\alpha. \tag{3.8}$$

Similarly,

$$\left\| \int_0^t \left((t+r-\tau)^{\alpha-1} - (t-\tau)^{\alpha-1} \right) g(\xi(\tau))\, d\tau \right\| \leq \frac{M}{\alpha} \left(-(t+r)^\alpha + r^\alpha + t^\alpha \right). \tag{3.9}$$

From (3.8) and (3.9) leads to

$$\left\| \mathcal{T}_f\xi(t+r) - \mathcal{T}_f\xi(t) \right\| \leq \|f(t+r) - f(t)\| + \frac{M}{\alpha\Gamma(\alpha)} \left(-(t+r)^\alpha + t^\alpha \right) + \frac{2M}{\alpha\Gamma(\alpha)} r^\alpha,$$

which shows that, when $t \in [0,T)$, $\left\| \mathcal{T}_f\xi(t+r) - \mathcal{T}_f\xi(t) \right\| \to 0$ as $r \to 0$. For the case $t = T$, by the same arguments as above, we also obtain

$$\left\| \mathcal{T}_f \xi (T - r) - \mathcal{T}_f \xi (T) \right\| \to 0 \quad \text{as } r \to 0.$$

Thus, the mapping $t \mapsto (\mathcal{T}_f \xi)(t)$ is continuous on $[0, T]$. This means that the operator $\mathcal{T}_f$ is well-defined from $C([0, T], \mathbb{R}^d)$ to $C([0, T], \mathbb{R}^d)$. $\qquad\square$

Lemma 3.3 *Assume that the vector field g satisfies the global Lipschitz condition in Assumption 3.1. Then, for any $T > 0$ and for each $f \in C([0, T], \mathbb{R}^d)$, the integral equation (3.5) has a unique solution $x_f(\cdot)$ on the interval $[0, T]$.*

Proof Since the proof follows from the Banach Contraction Mapping Theorem, it remains to show that the contraction property holds in the weighted norm (3.7). Let $\xi, \chi \in C([0, T], \mathbb{R}^d)$. Then, for each $t \in [0, T]$,

$$(\mathcal{T}_f \xi)(t) = f(t) + \frac{1}{\Gamma(\alpha)} \int_0^t (t - \tau)^{\alpha - 1} g(\xi(\tau)) \, d\tau,$$

$$(\mathcal{T}_f \chi)(t) = f(t) + \frac{1}{\Gamma(\alpha)} \int_0^t (t - s)^{\alpha - 1} g(\chi(s)) ds.$$

Then,

$$\left\| (\mathcal{T}_f \xi)(t) - (\mathcal{T}_f \chi)(t) \right\| \le \frac{1}{\Gamma(\alpha)} \int_0^t (t - s)^{\alpha - 1} \left\| g(\xi(s)) - g(\chi(s)) \right\| ds$$

$$\le \frac{L}{\Gamma(\alpha)} \int_0^t (t - s)^{\alpha - 1} \left\| \xi(s) - \chi(s) \right\| ds, \ \forall t \in [0, T].$$

By definition of the weighted norm $\| \cdot \|_\gamma$ and by the inequality (1.14),

$$\left\| (\mathcal{T}_f \xi)(t) - (\mathcal{T}_f \chi)(t) \right\| \le \frac{L}{\Gamma(\alpha)} \int_0^t (t - s)^{\alpha - 1} E_\alpha(\gamma s^\alpha) \, ds \, \| \xi - \chi \|_\gamma.$$

$$\le \frac{L}{\gamma} E_\alpha(\gamma t^\alpha) \| \xi - \chi \|_\gamma, \ \forall t \in [0, T].$$

This gives

$$\frac{\left\| (\mathcal{T}_f \xi)(t) - (\mathcal{T}_f \chi)(t) \right\|}{E_\alpha(\gamma t^\alpha)} \le \frac{L}{\gamma} \| \xi - \chi \|_\gamma, \ \forall t \in [0, T],$$

and hence

$$\left\| \mathcal{T}_f \xi - \mathcal{T}_f \chi \right\|_\gamma \le \frac{L}{\gamma} \| \xi - \chi \|_\gamma.$$

The choice $\gamma > L$ makes the operator $\mathcal{T}_f$ a contraction on the Banach space $(C([0, T], \mathbb{R}^d), \| \cdot \|_\gamma)$. Its unique fixed point gives the unique solution $x_f(\cdot)$ of (3.5) in $C([0, T], \mathbb{R}^d)$. This competes the proof of Lemma 3.3. $\qquad\square$

The solutions also depend continuously on the input function f, which will now be shown. This does not require the weighted norm (3.7), but will use a version of Gronwall's lemma, see Lemma 2.4.

Lemma 3.4 *Assume that the vector field g satisfies the global Lipschitz condition in Assumption 3.1. Then for any $T > 0$ and for each $f \in C([0, T], \mathbb{R}^d)$ the unique solution $x_f(\cdot)$ of the integral equation (3.5) depends continuously on f in the supremum norm.*

Proof Let $x_f, y_h \in \in C([0, T], \mathbb{R}^d)$ be the unique solutions of (3.5) corresponding to the inputs $f, h \in C([0, T], \mathbb{R}^d)$. Then,

$$x_f(t) - y_h(t) = f(t) - h(t) + \frac{1}{\Gamma(\alpha)} \int_0^t (t - s)^{\alpha-1} (g(x_f(s)) - g(y_h(s))) ds, \quad \forall t \in [0, T].$$

Thus,

$$\|x_f(t) - y_h(t)\| \le \|f(t) - h(t)\| + \frac{L}{\Gamma(\alpha)} \int_0^t (t - s)^{\alpha-1} \|x_f(s) - y_h(s)\| \, ds$$

$$\le \|f - h\|_\infty + \frac{L}{\Gamma(\alpha)} \int_0^t (t - s)^{\alpha-1} \|x_f(s) - y_h(s)\| \, ds, \quad \forall t \in [0, T].$$

By the fractional Gronwall Lemma 2.4, for all $t \in [0, T]$,

$$\left\| x_f(t) - y_h(t) \right\| \le \|f - h\|_\infty E_\alpha(Lt^\alpha).$$

Taking the supremum norm on $C([0, T], \mathbb{R}^d)$ then gives

$$\left\| x_f - y_h \right\|_\infty \le \|f - h\|_\infty E_\alpha(LT^\alpha).$$

The proof is complete. $\qquad \square$

The proof of Theorem 3.2 is obtained by combining Lemmas 3.2, 3.3 and 3.4.

3.2 Local Existence Results

Many interesting examples of Caputo FDEs involve continuously differentiable vector fields, which thus satisfy a local Lipschitz condition, but do not necessarily satisfy a global Lipschitz condition. In such cases only the local existence of a solution can be guaranteed. This also holds if the vector field is only continuous.

The following result is essentially a counterpart of the Peano theorem for ODEs.

Theorem 3.3 *Let $x_0 \in \mathbb{R}^d$, $K > 0$ and suppose that $g : G \to \mathbb{R}^d$ is a continuous, where*

$$G := \{(t, x) : 0 \le t \le T, \ \|y - x_0\| \le K\}.$$

Then there exists $T_b(x_0) \in (0, T)$ such that the IVP (3.1)–(3.2) has a solution $\varphi(\cdot, x_0) \in C([0, T_b(x_0)], \mathbb{R}^d)$. Moreover, $(t, \varphi(t, x_0)) \in G$ for all $0 \le t \le T_b(x_0)$.

Proof The proof uses the same arguments as in the proof of [2, Theorem 6.1]. $\square$

When the vector field satisfies a local Lipschitz condition then the existence of a unique solution on a maximal interval of existence can be established.

Theorem 3.4 *In addition to the assumptions of Theorem 3.3 assume that the function g is Lipschitz continuous on G. Then there exists (a maximal time) $T_b(x_0) \in (0, T]$ such that the IVP (3.1)–(3.2) has a unique solution $\varphi(\cdot, x_0) \in C([0, T_b(x_0)], \mathbb{R}^d)$. Moreover,*

$$(t, \varphi(t, x_0)) \in G \quad and \quad (T_b(x_0), \varphi(T_b(x_0), x_0)) \in \partial G$$

for any $0 \le t \le T_b(x_0)$, i.e.,

$$either \quad T_b(x_0) = T \quad or \quad T_b(x_0) < T \ with \quad \|\varphi(T_b(x_0), x_0) - x_0\| = K.$$

Proof The proof is follows directly from [5, Proposition 4.6]. $\square$

Remark 3.1 If the vector field satisfies a dissipativity condition as when as being locally Lipschitz, then the solutions exist can be extended to all time $t \ge 0$, see Theorems 6.1 and 9.1.

3.3 Hölder Continuity of the Solutions

Consider the integral equation (3.5), i.e.,

$$x(t) = f(t) + \frac{1}{\Gamma(\alpha)} \int_0^t (t - s)^{\alpha-1} g(x(s)) ds, \ t \ge 0.$$

It will be shown that the solutions of this equation are Hölder continuous with exponent α when the function f is Hölder continuous with exponent α.

Theorem 3.5 *Suppose that the input function f is Hölder continuous with exponent α. Then the solution of the integral equation (3.5) is Hölder continuous with exponent α.*

Proof The proof is similar to the first part of that of Lemma 3.2. Then, for $t, t + r \in [0, T]$ with $r > 0$,

$$x_f(t + r) - x_f(t) = f(t + r) - f(t) + \frac{1}{\Gamma(\alpha)} \int_0^{t+r} (t + r - \tau)^{\alpha-1} g(x_f(\tau)) \, d\tau$$

$$- \frac{1}{\Gamma(\alpha)} \int_0^t (t-\tau)^{\alpha-1} g(x_f(\tau)) \, d\tau$$

$$= f(t+r) - f(t) + \frac{1}{\Gamma(\alpha)} \int_t^{t+r} (t+r-\tau)^{\alpha-1} g(x_f(\tau)) \, d\tau$$

$$+ \frac{1}{\Gamma(\alpha)} \int_0^t \Big((t+r-\tau)^{\alpha-1} - (t-\tau)^{\alpha-1} \Big) g(x_f(\tau)) \, d\tau.$$

Let $M := \max_{t \in [0,T]} \| g(x_f(t)) \| < \infty$. Then,

$$\left\| \int_t^{t+r} (t+r-\tau)^{\alpha-1} g(x_f(\tau)) \, d\tau \right\| \leq \int_t^{t+r} (t+r-\tau)^{\alpha-1} \left\| g(x_f(\tau)) \right\| \, d\tau$$

$$\leq M \int_t^{t+r} (t+r-\tau)^{\alpha-1} d\tau \leq \frac{M}{\alpha} r^\alpha.$$

Similarly, for $t \in [0, T)$,

$$\left\| \int_0^t \Big((t+r-\tau)^{\alpha-1} - (t-\tau)^{\alpha-1} \Big) g(x(\tau)) \, d\tau \right\| \leq \frac{M}{\alpha} \Big(-(t+r)^\alpha + t^\alpha + r^\alpha \Big).$$

From this,

$$\| x_f(t+r) - x_f(t) \| \leq \| f(t+r) - f(t) \| + \frac{M}{\alpha \Gamma(\alpha)} \Big(-(t+r)^\alpha + t^\alpha \Big) + \frac{2M}{\alpha \Gamma(\alpha)} r^\alpha$$

$$\leq \| f \|_\alpha \, r^\alpha + \frac{2M}{\alpha \Gamma(\alpha)} r^\alpha,$$

where $\| f \|_\alpha$ is the Hölder norm of f. For the case $t = T$, by the same arguments, then

$$\| x_f(T-r) - x_f(T) \| \leq \| f(T-r) - f(T) \| + \frac{M}{\alpha \Gamma(\alpha)} \Big(-T^\alpha + (T-r)^\alpha \Big) + \frac{2M}{\alpha \Gamma(\alpha)} r^\alpha$$

$$\leq \| f \|_\alpha \, r^\alpha + \frac{2M}{\alpha \Gamma(\alpha)} r^\alpha.$$

Thus, the solution $x_f(\cdot)$ is Hölder continuous with exponent α. $\qquad\square$

In the special case with $f(t) \equiv x_0$ this says that the solution $\varphi(\cdot, x_0)$ of the Caputo FDE (3.1) is Hölder continuous with exponent α.

3.4 Continuous Dependence on the Fractional Index

The solutions of Caputo FDEs also depend continuously on the fractional exponent α. The main ingredient of the proof is to use the suitable weighted norm as in Theorem 3.2.

To simplify the notation denote by the kernel function depending on the fractional exponent $\alpha \in (0, 1)$ by

$$a_\alpha(t, s) := \frac{1}{\Gamma(\alpha)}(t - s)^{\alpha-1}, \qquad 0 \le s < t.$$

Thanks to Theorem 3.2 for any $T > 0$ and any $f \in C([0, T], \mathbb{R}^d)$, the integral equation

$$x(t) = f(t) + \int_0^t a_\alpha(t, s)g(x(s))ds, \ \forall t \in [0, T], \tag{3.10}$$

has a unique solution on $[0, T]$ denoted by $x_\alpha(\cdot, f)$.

Theorem 3.6 *Assume that the vector field g satisfies the global Lipschitz condition in Assumption 3.1. Then for any $T > 0$ and for each $f \in C([0, T], \mathbb{R}^d)$, the unique solution $x_\alpha(\cdot, f)$ of the integral equation (3.10) depends continuously on α in the supremum norm.*

Proof Fix an arbitrary $\alpha \in (0, 1)$ and let $\beta \in [\frac{\alpha}{2}, 1]$ be arbitrary. By definition of the solution

$$x_\alpha(t, f) - x_\beta(t, f) = \int_0^t a_\alpha(t, s)g(x_\alpha(s, f)) \, ds - \int_0^t a_\beta(t, s)g(x_\beta(s, f)) \, ds.$$

Then, by Assumption 3.1,

$$\|x_\alpha(t, f) - x_\beta(t, f)\| \le \left\| \int_0^t a_\alpha(t, s) \left(g(x_\alpha(s, f)) - g(x_\beta(s, f))\right) ds \right\|$$

$$+ \left\| \int_0^t \left(a_\alpha(t, s) - a_\beta(t, s)\right) g(x_\beta(s, f)) ds \right\|$$

$$\le L \int_0^t a_\alpha(t, s)\|x_\alpha(s, f) - x_\beta(s, f)\| \, ds$$

$$+ M_\beta \int_0^t |a_\alpha(t, s) - a_\beta(t, s)| \, ds, \ \forall t \in [0, T],$$

where $M_\beta := \max_{0 \le s \le T} \|g(x_\beta(s, f))\|$.

Let $\gamma > 0$ be arbitrary and let $C([0, T], \mathbb{R}^d)$ be endowed with the weight norm $\|\cdot\|_\gamma$ defined by

$$\|x\|_\gamma := \sup_{t \in [0,T]} \frac{\|x(t)\|}{E_\alpha(\gamma t^\alpha)} \qquad \text{for all } x \in C([0, T], \mathbb{R}^d).$$

Then, by (1.14), the following estimates hold

$$\frac{\|x_\alpha(t,f)-x_\beta(t,f)\|}{E_\alpha(\gamma t^\alpha)} \leq L \int_0^t \frac{a_\alpha(t,s)E_\alpha(\gamma s^\alpha)ds}{E_\alpha(\gamma t^\alpha)} \|x_\alpha(\cdot,f)-x_\beta(\cdot,f)\|_\gamma$$

$$+M_\beta \int_0^t |a_\alpha(t,s)-a_\beta(t,s)|\,ds$$

$$\leq \frac{L}{\gamma}\|x_\alpha(\cdot,f)-x_\beta(\cdot,f)\|_\gamma$$

$$+M_\beta \left| \int_0^T \frac{1}{\Gamma(\alpha)}\frac{1}{(T-s)^{1-\alpha}}\,ds - \frac{1}{\Gamma(\beta)}\int_0^T \frac{1}{(T-s)^{1-\beta}}\,ds \right|$$

$$\leq \frac{L}{\gamma}\|x_\alpha(\cdot,f)-x_\beta(\cdot,f)\|_\gamma$$

$$+M_\beta \left| \frac{1}{\alpha\Gamma(\alpha)}T^\alpha - \frac{1}{\beta\Gamma(\beta)}T^\beta \right|.$$

Consequently, a fixed choice of γ satisfying that $\gamma > L$, leads to

$$\|x_\alpha(\cdot,f)-x_\beta(\cdot,f)\|_\gamma \leq \frac{M_\beta}{1-\frac{L}{\gamma}} \left| \frac{1}{\alpha\Gamma(\alpha)}T^\alpha - \frac{1}{\beta\Gamma(\beta)}T^\beta \right|$$

$$\leq \frac{\sup_{\beta\in[\frac{\alpha}{2},1]} M_\beta}{1-\frac{L}{\gamma}} \left| \frac{1}{\alpha\Gamma(\alpha)}T^\alpha - \frac{1}{\beta\Gamma(\beta)}T^\beta \right|.$$

Hence, $\|x_\alpha(\cdot,f)-x_\beta(\cdot,f)\|_\gamma \to 0$ as $\beta \to \alpha$, provided the supremum is finite. Thus, to conclude the proof, it remains to show that $\sup_{\beta\in[\frac{\alpha}{2},1]} M_\beta < \infty$. By definition of M_β and Assumption 3.1

$$M_\beta \leq \|g(0)\| + L \sup_{0\leq t\leq T} \|x_\beta(t,f)\|. \tag{3.11}$$

Then it follows from

$$x_\beta(t,f) = f(t) + \int_0^t a_\beta(t,s)g(x_\beta(s,f))\,ds$$

that

$$\|x_\beta(t,f)\| \leq \|f\|_\infty + \|g(0)\| \int_0^t a_\beta(t,s)\,ds + L \int_0^t a_\beta(t,s)\|x_\beta(s,f)\|\,ds.$$

A direct computation yields $\int_0^t a_\beta(t,s)\,ds = \frac{t^\beta}{\beta\Gamma(\beta)}$, so

$$\|x_\beta(t,f)\| \leq \|f\|_\infty + \frac{\|g(0)\|T^\beta}{\beta\Gamma(\beta)} + L \int_0^t a_\beta(t,s)\|x_\beta(s,f)\|\,ds.$$

From this, the fractional Gronwall inequality gives

$$\|x_\beta(t, f)\| \le \left(\|f\|_\infty + \frac{\|g(0)\|T^\beta}{\beta\Gamma(\beta)} \right) E_\beta(LT^\beta),$$

which together with (3.11) implies that

$$\sup_{\beta\in[\frac{\alpha}{2},1]} M_\beta \le \|g(0)\| + L \left(\|f\|_\infty + \|g(0)\| \max_{\beta\in[\frac{\alpha}{2},1]} \frac{T^\beta}{\beta\Gamma(\beta)} \right) \max_{\beta\in[\frac{\alpha}{2},1]} E_\beta(LT^\beta) < \infty.$$

This completes the proof. $\square$

3.5   Nonlinear Variation of Constants Formula

Consider a scalar Caputo FDE

$$^C D_{0+}^\alpha x(t) = g(x(t)), \quad \forall t > 0, \tag{3.12}$$

where $g : \mathbb{R} \to \mathbb{R}$ satisfies the global Lipschitz condition

$$\|g(x) - g(y)\| \le L\|x - y\|, \quad x, y \in \mathbb{R}. \tag{3.13}$$

Lemma 3.5 (Variation of constants formula) *Suppose that the vector field g of the scalar Caputo FDE* (3.12) *satisfies the global Lipschitz condition* (3.13) *and has the form*

$$g(x) = \gamma x + n(x)$$

for some fixed $\gamma \in \mathbb{R} \setminus \{0\}$. Then, the solution $x(\cdot)$ of (3.12) *with the initial value $x(0) = x_0$ satisfies the formula*

$$x(t) = E_\alpha(\gamma t^\alpha)x_0 + \int_0^t (t - \tau)^{\alpha-1} E_{\alpha,\alpha}(\gamma(t - \tau)^\alpha)n(x(\tau))d\tau, \quad t \ge 0.$$

Proof By the Lipschitz condition (3.13) there exists a positive constant K such that

$$|n(0)| \le K, \quad |n(x) - n(y)| \le K|x - y|, \quad \forall x, y \in \mathbb{R}.$$

The Caputo FDE (3.12) can be rewritten as

$$^C D_{0+}^\alpha x(t) = \gamma x(t) + n(x(t)) = \gamma x(t) + n(x(t)) - n(0) + n(0), \quad \forall t > 0. \tag{3.14}$$

By Theorem 3.1, the Eq. (3.14) with the initial condition $x(0) = x_0$ has a unique solution which satisfies the Volterra integral equation

$$x(t) = x_0 + \frac{1}{\Gamma(\alpha)} \int_0^t (t - \tau)^{\alpha-1} \left[\gamma x(\tau) + n(x(\tau)) - n(0) \right] d\tau$$

$$+ \frac{1}{\Gamma(\alpha)} \int_0^t (t - \tau)^{\alpha-1} n(0) \, d\tau, \quad t \geq 0.$$

Hence,

$$|x(t)| \leq |x_0| + \frac{K t^\alpha}{\alpha \Gamma(\alpha)} + \frac{K + |\gamma|}{\Gamma(\alpha)} \int_0^t (t - \tau)^{\alpha-1} |x(\tau)| \, d\tau, \quad \forall t \geq 0.$$

This implies that

$$|x(t)| e^{-t} \leq \left(|x_0| + \frac{K t^\alpha}{\alpha \Gamma(\alpha)} \right) e^{-t} + \frac{K + |\gamma|}{\Gamma(\alpha)} \int_0^t (t - \tau)^{\alpha-1} |x(\tau)| e^{-\tau} \, d\tau, \quad \forall t \geq 0.$$

$$(3.15)$$

Define $v(t) = |x(t)| e^{-t}$ for $t \geq 0$ and $M := \sup_{t \geq 0} \left(|x_0| + \frac{K t^\alpha}{\alpha \Gamma(\alpha)} \right) e^{-t} < \infty$. Then, from (3.15),

$$v(t) \leq M + \frac{K + |\gamma|}{\Gamma(\alpha)} \int_0^t (t - \tau)^{\alpha-1} v(\tau) \, d\tau, \quad \forall t \geq 0.$$

Using Lemma 2.4 gives

$$v(t) \leq M E_\alpha((K + |\gamma|)t^\alpha), \quad t \geq 0.$$

Thus,

$$x(t) \leq M e^t E_\alpha((K + |\gamma|)t^\alpha), \quad t \geq 0.$$

This means the solution $x(\cdot)$ of the Eq. (3.14) with the initial value $x(0) = x_0$ is exponentially bounded on $[0, \infty)$. Hence, the Laplace transform (see [2, Chap. 7] and [6, Chap. 5]) can be applied to both sides of the Eq. (3.14) to give

$$s^\alpha \mathcal{L}\{x(t)\}(s) - s^{\alpha-1} x_0 = \gamma \mathcal{L}\{x(t)\}(s) + \mathcal{L}\{n(x(t))\}(s), \quad Re(s) > c,$$

for some positive constant c [2, Theorem 7.1, p. 134]. This implies that

$$\mathcal{L}\{x(t)\}(s) = \frac{s^{\alpha-1}}{s^\alpha - \gamma} x_0 + \frac{1}{s^\alpha - \gamma} \mathcal{L}\{n(x(t))\}(s), \quad Re(s) > \max\{c, |\gamma|^{1/\alpha}\}.$$

Taking the inverse Laplace transform then gives

$$x(t) = \mathcal{L}^{-1}\left\{ \frac{s^{\alpha-1}}{s^\alpha - \gamma} \right\}(t) x_0 + \mathcal{L}^{-1}\left\{ \frac{1}{s^\alpha - \gamma} \mathcal{L}\{n(x(t))\}(s) \right\}(t)$$

for all $t \geq 0$. The following formula [7, formula (1.80), p. 21] of Laplace transform of the Mittag-Leffler function holds:

$$\mathcal{L}\{t^{\beta-1}E_{\alpha,\beta}(\gamma t^{\alpha})\}(s) = \frac{s^{\alpha-\beta}}{s^{\alpha}-\gamma}, \quad \forall\, Re(s) > |\gamma|^{1/\alpha},$$

with $\beta = \alpha$ or $\beta = 1$. Using it and the properties of the Laplace transform [2, Theorem D.11, p. 231] gives

$$x(t) = E_{\alpha}(\gamma t^{\alpha})x_0 + \int_0^t (t-\tau)^{\alpha-1}E_{\alpha,\alpha}(\gamma(t-\tau)^{\alpha})n(x(\tau))\,d\tau, \quad t \geq 0.$$

This completes the proof. $\qquad\qquad\qquad\qquad\qquad\qquad\qquad\qquad\qquad\qquad\qquad\square$

Remark 3.2 A more general version of Lemma 3.5 where the vector field f depends on both the time variable t and the spatial variable x was introduced by Cong and Tuan [8, Lemma 3.1].

Remark 3.3 The proof of Lemma 3.5 can be easily adapted to the higher dimensional case with γ changed to a constant matrix and the functions x, g, n changed to vector functions, see [2, Remark 7.1, p. 135].

3.6   Time-Translation Invariance of Autonomous Solutions

Consider the IVP

$$\begin{cases} {}^{C}D_{0+}^{\alpha}x(t) = g(x(t)), & t > 0, \\ x(0) = x_0 \end{cases} \tag{3.16}$$

for an autonomous or the Caputo FDE, i.e., with the vector field g that does not depend explicitly on time. Here, by convention, the initial time $t_0 = 0$. Suppose that the solution of the IVP (3.16) is unique and denote it by $x(t; x_0)$.

In some modeling situations one may wish to consider an initial time $t_0 \neq 0$, i.e., the IVP

$${}^{C}D_{t_0+}^{\alpha}y(t) = g(y(t)), \quad y(t_0) = y_0, \qquad t \geq t_0, \tag{3.17}$$

and denote its solution by $y(t; t_0, y_0)$.

Then, as for the corresponding ODEs the solutions with the same initial value, i.e., $y_0 = x_0$, are *invariant under time translation*, i.e.,

$$y(\tau + t_0; t_0, x_0) = x(\tau; x_0) \qquad \text{for all } \tau \geq 0.$$

This says, essentially, that the solutions depend only on the elapsed time since starting and not on the actually values of time themselves.

For autonomous ODE case this is a direct consequence of the autonomous vector field and the uniqueness of solutions of the IVP. For autonomous Caputo FDEs it also makes use of the fact that the kernel

$$a_\alpha(t, s) = \frac{1}{\Gamma(\alpha)}(t - s)^{\alpha - 1}$$

in the integral equation representation depends only on the time difference $t - s$ and not on the actual values of t and s themselves.

Theorem 3.7 *The solutions of an autonomous Caputo FDE are invariant under time translation.*

Proof The solution $y(t) = y(t; t_0, x_0)$ of the IVP (3.17) satisfies the integral equation

$$y(t) = x_0 + \frac{1}{\Gamma(\alpha)} \int_{t_0}^{t} (t - s)^{\alpha - 1} g(y(s)) ds, \qquad t \geq t_0.$$

Thus, for $\tau \geq 0$,

$$y(\tau + t_0) = x_0 + \frac{1}{\Gamma(\alpha)} \int_{t_0}^{t_0 + \tau} (t_0 + \tau - s)^{\alpha - 1} g(y(s)) ds$$

$$= x_0 + \frac{1}{\Gamma(\alpha)} \int_{0}^{\tau} (t_0 + \tau - r - t_0)^{\alpha - 1} g(y(r + t_0)) dr \qquad (r := s - t_0)$$

$$= x_0 + \frac{1}{\Gamma(\alpha)} \int_{0}^{\tau} (\tau - r)^{\alpha - 1} g(y(r + t_0)) dr.$$

Hence, $\bar{x}(t) := y(t + t_0)$, $t \geq 0$, satisfies

$$\bar{x}(t) = x_0 + \frac{1}{\Gamma(\alpha)} \int_{0}^{t} (t - s)^{\alpha - 1} g(\bar{x}(s)) ds, \qquad t \geq 0,$$

which means $\bar{x}(t) = x(t, x_0)$ since the solution of the IVP (3.16) is unique. $\square$

In view of Theorem 3.7, there is no loss of generality in restricting the initial time to $t_0 = 0$.

3.7 Boundary Value Problems

Boundary value problems arise in many applications including the investigation of dynamical systems properties of Caputo fractional differential equations.

A first order fractional boundary value problem (BVP) for a Caputo fractional differential equation (FDE) of order $\alpha \in (0, 1)$ in $\mathbb{R}^d$ has the form

$$^{C}D_{0+}^{\alpha}x(t) = g(x(t)), \tag{3.18}$$

$$ax(0) + bx(T) = c, \tag{3.19}$$

where $g : \mathbb{R}^d \to \mathbb{R}^d$ is (at least) continuous, a, b are real constants with $a + b \neq 0$ and $c \in \mathbb{R}^d$ is a constant vector.

This formulation includes initial value problems ($a = 1$, $b = 0$), terminal value problems ($a = 0$, $b = 1$) and anti-periodic problems ($a = b = 1$, $c = 0$).

The following auxiliary lemma is needed to establish the existence of a solution of such a BVP.

Lemma 3.6 [9, Lemma 6] *Let $\alpha \in (0, 1)$ and let $h : [0, T] \to \mathbb{R}^d$ be continuous. Then a function $x : [0, T] \to \mathbb{R}^d$ is a solution of the fractional integral equation*

$$x(t) = \frac{1}{\Gamma(\alpha)} \int_0^t (t - s)^{\alpha - 1} h(s)ds - \frac{1}{a + b} \left[\frac{b}{\Gamma(\alpha)} \int_0^T (T - s)^{\alpha - 1} h(s)ds - c \right]$$

if and only if it is a solution on the interval $[0, T]$ of the fractional boundary value problem

$$^{C}D_{0+}^{\alpha}x(t) = h(t), \tag{3.20}$$

$$ax(0) + bx(T) = c. \tag{3.21}$$

Two existence theorems will be presented here for solutions of the BVP (3.20)–(3.21). Both involve fixed points of the operator

$$\mathfrak{F} : C([0, T], \mathbb{R}^d) \to C([0, T], \mathbb{R}^d)$$

defined by

$$(\mathfrak{F}\xi)(t) = \frac{1}{\Gamma(\alpha)} \int_0^t (t - s)^{\alpha - 1} g(\xi(s))ds$$

$$- \frac{1}{a + b} \left[\frac{b}{\Gamma(\alpha)} \int_0^T (T - s)^{\alpha - 1} g(\xi(s))ds - c \right], \quad \forall t \in [0, T] \tag{3.22}$$

for $\xi \in C([0, T], \mathbb{R}^d)$.

3.7.1   An Existence Theorem

The following existence theorem for a BVP requires a restriction on the time T under consideration and that the vector field $g : \mathbb{R}^d \to \mathbb{R}^d$ is globally Lipschitz continuous.

Theorem 3.8 [9, Theorem 7] *Assume that the vector field g satisfies the global Lipschitz condition*

$$\|g(x) - g(y)\| \leq L\|x - y\|, \quad x, y \in \mathbb{R}^d. \tag{3.23}$$

If

$$\left(1 + \frac{|b|}{|a+b|}\right)\frac{LT^\alpha}{\alpha\Gamma(\alpha)} < 1, \tag{3.24}$$

then the BVP (3.20)–(3.21) has a solution on the interval $[0, T]$.

Proof The Banach contraction principle will be used to show that the operator $\mathfrak{F}$: $C([0, T], \mathbb{R}^d) \to C([0, T], \mathbb{R}^d)$ defined by (3.22) has a fixed point in the space $C([0, T], \mathbb{R}^d)$ with the supremum norm. Clearly, such a fixed point is a solution of the BVP (3.20)–(3.21).

The well-definedness of $\mathfrak{F}$ follows in the same way as in Lemma 3.2 for the corresponding initial value problem.

It remains to show that $\mathfrak{F}$ is a contraction. Let $\xi, \chi \in C([0, T], \mathbb{R}^d)$, so for each $t \in [0, T]$

$$(\mathfrak{F}\xi)(t) = \frac{1}{\Gamma(\alpha)}\int_0^t (t-s)^{\alpha-1}g(\xi(s))\,ds + \frac{1}{a+b}\left[\frac{b}{\Gamma(\alpha)}\int_0^T (T-s)^{\alpha-1}g(\xi(s))ds - c\right],$$

$$(\mathfrak{F}\chi)(t) = \frac{1}{\Gamma(\alpha)}\int_0^t (t-s)^{\alpha-1}g(\chi(s))ds + \frac{1}{a+b}\left[\frac{b}{\Gamma(\alpha)}\int_0^T (T-s)^{\alpha-1}g(\chi(s))ds - c\right].$$

Then

$$\|(\mathfrak{F}\xi)(t) - (\mathfrak{F}\chi)(t)\| \leq \frac{1}{\Gamma(\alpha)}\int_0^t (t-s)^{\alpha-1}\|g(\xi(s)) - g(\chi(s))\|\,ds$$

$$+\frac{|b|}{|a+b|}\frac{1}{\Gamma(\alpha)}\int_0^T (T-s)^{\alpha-1}\|g(\xi(s)) - g(\chi(s))\|\,ds$$

$$\leq \frac{L}{\Gamma(\alpha)}\int_0^t (t-s)^{\alpha-1}\|\xi(s) - \chi(s)\|\,ds$$

$$+\frac{|b|}{|a+b|}\frac{L}{\Gamma(\alpha)}\int_0^T (T-s)^{\alpha-1}\|\xi(s) - \chi(s)\|\,ds$$

$$\leq \frac{L\|\xi - \chi\|_\infty}{\Gamma(\alpha)}\int_0^t (t-s)^{\alpha-1}ds$$

$$+\frac{|b|}{|a+b|}\frac{L\|\xi - \chi\|_\infty}{\Gamma(\alpha)}\int_0^T (T-s)^{\alpha-1}ds$$

for $t \in [0, T]$. Hence

$$\|\mathfrak{F}\xi - \mathfrak{F}\chi\|_\infty \leq \left(1 + \frac{|b|}{|a+b|}\right) \frac{LT^\alpha}{\alpha\Gamma(\alpha)} \|\xi - \chi\|_\infty ,$$

so $\mathfrak{F}$ is a contraction on the Banach space $(C([0, T], \mathbb{R}^d), \|\cdot\|_\infty)$ under the restriction (3.24) on T. Its unique fixed point is a solution of the BVP (3.20)–(3.21). This competes the proof of Theorem 3.8. $\qquad\square$

3.7.2 An Existence Theorem Without a Restriction on the End-Time

The existence of a solution to the BVP (3.20)–(3.21) can be shown without a restriction on the end-time T, but it requires the vector field function g of the Caputo FDE to be uniformly bounded. This solution need not be unique. The proof uses the fixed point theorem of Schaefer [10]. Recall that a compact (or completely continuous) function maps bounded sets into pre-compact sets.

Theorem 3.9 (Schaefer's fixed point theorem) [11, Theorem 10.1] *Let X be a Banach space and $f : X \to X$ be continuous and compact. Assume that the set*

$$\mathfrak{S} = \{x \in X \; : \; x = \lambda f(x) \text{ for some } \lambda \in [0, 1]\}$$

is bounded. Then f has a fixed point $\bar{x} \in X$.

Note that the set $\mathfrak{S}$ is nonempty since $x = 0 \in \mathfrak{S}$ for $\lambda = 0$.

Theorem 3.10 [9, Theorem 8] *Assume that the vector field $g : \mathbb{R}^d \to \mathbb{R}^d$ is continuous and that there is an $M > 0$ such that*

$$\|g(x)\| \leq M \text{ for all } x \in \mathbb{R}^d. \tag{3.25}$$

Then the BVP (3.20)–(3.21) has a solution on the interval $[0, T]$.

Proof Consider the operator $\mathfrak{F} : C([0, T], \mathbb{R}^d) \to C([0, T], \mathbb{R}^d)$ defined by (3.22). As noted in the proof of Theorem 3.10 the operator $\mathfrak{F}$ is well-defined. Schaefer's fixed point theorem, Theorem 3.9, will be used to show that $\mathfrak{F}$ has a fixed point. The proof is divided into four steps.

Step 1: Continuity of $\mathfrak{F}$.

Consider a sequence $\xi_n \to \xi$ in $C([0, T], \mathbb{R}^d)$ in the supremum norm. Then, as in the proof of Theorem 3.10,

$$\|(\mathfrak{F}\xi_n)(t) - (\mathfrak{F}\xi)(t)\| \leq \frac{1}{\Gamma(\alpha)} \int_0^t (t-s)^{\alpha-1} \|g(\xi_n(s)) - g(\xi(s))\| \, ds$$

$$+ \frac{|b|}{|a+b|} \frac{1}{\Gamma(\alpha)} \int_0^T (T-s)^{\alpha-1} \|g(\xi_n(s)) - g(\xi(s))\| \, ds$$

$$\leq \frac{1}{\Gamma(\alpha)} \int_0^t (t-s)^{\alpha-1} ds \sup_{s\in[0,T]} \|g(\xi_n(s)) - g(\xi(s))\|$$

$$+ \frac{|b|}{|a+b|} \frac{1}{\Gamma(\alpha)} \int_0^T (T-s)^{\alpha-1} ds \sup_{s\in[0,T]} \|g(\xi_n(s)) - g(\xi(s))\|$$

$$\leq \frac{\|g(\xi_n) - g(\xi)\|_\infty}{\Gamma(\alpha)} \left(\int_0^t (t-s)^{\alpha-1} ds + \frac{|b|}{|a+b|} \int_0^T (T-s)^{\alpha-1} ds \right)$$

for each $t \in [0, T]$. Hence

$$\|\mathfrak{F}\xi_n - \mathfrak{F}\xi\|_\infty \leq \left(1 + \frac{|b|}{|a+b|} \right) \frac{T^\alpha}{\alpha\Gamma(\alpha)} \|g(\xi_n) - g(\xi)\|_\infty \to 0 \text{ as } n \to \infty,$$

since g is a continuous function.

Step 2: $\mathfrak{F}$ maps bounded sets into bounded sets of $C([0, T], \mathbb{R}^d)$.

It suffices to show that for any $\eta > 0$ there exists a positive constant K such that $\|\mathfrak{F}\xi\| \leq K$ for all $\xi \in B_\eta := \{\chi \in C([0, T], \mathbb{R}^d) : \|\chi\|_\infty \leq \eta\}$.

By the assumption (3.25), for each $t \in [0, T]$,

$$\|(\mathfrak{F}\xi)(t)\| \leq \frac{1}{\Gamma(\alpha)} \int_0^t (t-s)^{\alpha-1} \|g(\xi(s))\| \, ds$$

$$+ \frac{|b|}{|a+b|} \frac{1}{\Gamma(\alpha)} \int_0^T (T-s)^{\alpha-1} \|g(\xi(s))\| \, ds + \frac{|c|}{|a+b|}$$

$$\leq \frac{M}{\Gamma(\alpha)} \int_0^t (t-s)^{\alpha-1} ds + \frac{M|b|}{|a+b|} \frac{1}{\Gamma(\alpha)} \int_0^T (T-s)^{\alpha-1} ds + \frac{|c|}{|a+b|}$$

$$\leq \frac{M}{\alpha\Gamma(\alpha)} T^\alpha + \frac{M|b|}{|a+b|} \frac{1}{\alpha\Gamma(\alpha)} T^\alpha + \frac{|c|}{|a+b|}.$$

Hence

$$\|\mathfrak{F}\xi\|_\infty \leq \frac{MT^\alpha}{\alpha\Gamma(\alpha)} \left(1 + \frac{|b|}{|a+b|} \right) + \frac{|c|}{|a+b|} =: K.$$

Step 3: $\mathfrak{F}$ maps bounded sets into equicontinuous sets of $C([0, T], \mathbb{R}^d)$.

Let $t_1, t_2 \in [0, T]$ with $t_1 < t_2$ and let $\xi \in B_\eta$, where B_η is the ball defined in **Step 2**. Then

$$\|(\mathfrak{F}\xi)(t_2) - (\mathfrak{F}\xi)(t_1)\| \leq \frac{1}{\Gamma(\alpha)} \int_0^{t_1} \left(-(t_2-s)^{\alpha-1} + (t_1-s)^{\alpha-1} \right) \|g(\xi(s))\| \, ds$$

$$+ \frac{1}{\Gamma(\alpha)} \int_{t_1}^{t_2} (t_2-s)^{\alpha-1} \|g(\xi(s))\| \, ds$$

$$\leq \frac{M}{\Gamma(\alpha)} \int_0^{t_1} \left(-(t_2-s)^{\alpha-1} + (t_1-s)^{\alpha-1} \right) ds + \frac{M}{\Gamma(\alpha)} \int_{t_1}^{t_2} (t_2-s)^{\alpha-1} ds$$

$$\le \frac{M}{\alpha \Gamma(\alpha)}\left((t_2 - t_1)^\alpha - t_2^\alpha + t_1^\alpha\right) + \frac{M}{\alpha \Gamma(\alpha)}(t_2 - t_1)^\alpha$$

$$\le \frac{2M}{\alpha \Gamma(\alpha)}(t_2 - t_1)^\alpha + \frac{M}{\alpha \Gamma(\alpha)}\left(-t_2^\alpha + t_1^\alpha\right)$$

$$\to 0 \text{ as } t_1 \to t_2.$$

It follows from the steps 1 to 3 and the Arzelà-Ascoli Theorem that the operator $\mathfrak{F} : C([0, T], \mathbb{R}^d) \to C([0, T], \mathbb{R}^d)$ is continuous and compact.

Step 4: A priori bound.

It remains to show that the set

$$\mathfrak{E} = \left\{\xi \in C([0, T], \mathbb{R}^d) \ : \ \xi = \lambda \mathfrak{F}\xi \text{ for some } \lambda \in [0, 1]\right\}$$

is bounded.

Let $\xi \in \mathfrak{E}$. Then $\xi = \lambda \mathfrak{F}\xi$ for some $\lambda \in [0, 1]$. Thus, for each $t \in [0, T]$,

$$\xi(t) = \lambda \left[\frac{1}{\Gamma(\alpha)} \int_0^t (t - s)^{\alpha-1} g(\xi(s)) ds \right.$$

$$\left. - \frac{1}{a + b}\left(\frac{b}{\Gamma(\alpha)} \int_0^T (T - s)^{\alpha-1} g(\xi(s)) ds - c\right)\right].$$

Then, by the uniform bound (3.25), for each $t \in [0, T]$

$$\|(\mathfrak{F}\xi)(t)\| \le \frac{1}{\Gamma(\alpha)} \int_0^t (t - s)^{\alpha-1} \|g(\xi(s))\| \, ds$$

$$+ \frac{|b|}{|a + b|} \frac{1}{\Gamma(\alpha)} \int_0^T (T - s)^{\alpha-1} \|g(\xi(s))\| \, ds + \frac{|c|}{|a + b|}$$

$$\le \frac{M}{\Gamma(\alpha)} \int_0^t (t - s)^{\alpha-1} ds + \frac{M|b|}{|a + b|} \frac{1}{\Gamma(\alpha)} \int_0^T (T - s)^{\alpha-1} ds + \frac{|c|}{|a + b|}$$

$$\le \frac{M}{\alpha \Gamma(\alpha)} T^\alpha + \frac{M|b|}{|a + b|} \frac{1}{\alpha \Gamma(\alpha)} T^\alpha + \frac{|c|}{|a + b|}$$

Hence

$$\|\mathfrak{F}\xi\|_\infty \le \frac{M T^\alpha}{\alpha \Gamma(\alpha)}\left(1 + \frac{|b|}{|a + b|}\right) + \frac{|c|}{|a + b|} < \infty,$$

so the set $\mathfrak{E}$ is bounded in $C([0, T], \mathbb{R}^d)$ in the supremum norm.

Schaefer's fixed point theorem can now be applied to conclude that there is a fixed point $\xi^* = \mathfrak{F}\xi^* \in \mathfrak{E}$, which is a solution of the BVP (3.20)–(3.21). This competes the proof of Theorem 3.10. $\qquad\square$

Remark 3.4 Note that the existence of a fixed point of the operator $\mathfrak{F}$ does not mean that the BVP has a unique solution. See the counterexample with a 2-dimensional linear Caputo fractional differential equation in Theorem 7.5.

3.8 Endnotes

The existence and uniqueness theory for initial vale problems is based on Diethelm [2]. The idea of the weighted norm is taken from Cong and Tuan [3].

The boundary value results are based on Benchohra, Hamani and and Ntouyas [9]. The proofs of the above theorems were given in the scalar case in [9], but remain valid in the vector setting here. See also Diethelm [2, Sect. 6.5].

A proof of Schaefer's Fixed Point Theorem is given by Pata [11, Theorem 10.1], where its relationship to other fixed point theorems is discussed.

References

1. Miller, R.K., Sell, G.R.: Volterra integral equations and topological dynamics. Memoir Amer. Math. Soc. **102** (1970)
2. Diethelm, K.: The analysis of fractional differential equations, Springer Lecture Notes in Mathematics, vol. 2004. Springer, Heidelberg (2010)
3. Cong, N.D., Tuan, H.T.: Existence, uniqueness and exponential boundedness of global solutions to delay fractional differential equations. Mediterr. J. Math. **14**, 1–12 (2017)
4. Doan, T.S., Kloeden, P.E.: Semi-dynamical systems generated by autonomous Caputo fractional differential equations. Vietnam J. Math. **49**, 1305–1315 (2021). https://doi.org/10.1007/s10013-020-00464-6
5. Li, L., Liu, J.: A generalized definition of Caputo derivatives and its applications to fractional ODEs. SIAM J. Math. Anal. **50**(3), 2867–2900 (2018)
6. Kilbas, A.A., Srivastava, H.M., Trujillo, J.J.: Theory and applications of fractional differential equations. North-Holland Mathematics Studies 204, Elsevier Science B.V., Amsterdam (2006). MR2218073
7. Podlubny, I.: Fractional differential equations. An introduction to fractional derivatives, fractional differential equations, to methods of their solution and some of their applications. Academic Press, Inc., San Diego, CA (1999)
8. Cong, N.D., Tuan, H.T.: Generation of nonlocal dynamical systems by fractional differential equations. J. Integral Equations Appl. **29**, 585–608 (2017). https://doi.org/10.1216/JIE-2017-29-4-585
9. Benchohra, M., Hamani, S., Ntouyas, S.K.: Boundary value problems for differential equations with fractional order. Surv. Math. Appl. **3**, 1–12 (2008)
10. Schaefer, H.: Über die Methode der a priori-Schranken. Math. Annalen **129**, 415–416 (1955)
11. Pata, V.: Fixed point theorems and applications. Springer Nature, Cham (2019)

Chapter 4
Linear Systems of Caputo FDEs

Abstract The content of this chapter consists of two parts. First, the stability and asymptotic stability of linear fractional differential equations with constant coefficients in arbitrary finite-dimensional spaces are discussed based on the fundamental solution expressed by the matrix-valued Mittag–Leffler function. Next, linear non-autonomous fractional differential systems are investigated with the fractional Lyapunov exponent. Some fundamental properties of this exponent are described and an application to the stability analysis of such systems is presented.

Keywords Caputo fractional-order linear systems · Mittag-Leffler functions · Classical Lyapunov exponent · Fractional Lyapunov exponent · Stability · Asymptotic stability

Consider a linear Caputo FDE

$$^{C}D_{0+}^{\alpha}x(t) = Ax(t), \ \forall t > 0, \tag{4.1}$$

where A is a $d \times d$ matrix.

Since the vector field $g(x) = Ax$ is linear and hence globally Lipschitz, the existence and uniqueness of solutions of IVPs holds by Theorem 3.2.

The solutions of the linear Caputo FDE (4.1) with the initial condition $x(0) = x_0 \in \mathbb{R}^d$ can be represented as matrix-valued Mittag-Leffler functions (see Chap. 1), namely

$$x(t) = E_{\alpha}(At^{\alpha})x_0, \ t \geq 0. \tag{4.2}$$

Moreover, the solution of the inhomogeneous linear vector-valued Caputo FDE

$$^{C}D_{0+}^{\alpha}x(t) = Ax(t) + h(t), \ t > 0$$

with the initial value $x(0) = x_0 \in \mathbb{R}^d$, where $h := [0, \infty) \to \mathbb{R}^d$ continuous, is given by

© The Author(s), under exclusive license to Springer Nature Switzerland AG 2025
T. S. Doan et al., *Attractors of Caputo Fractional Differential Equations*,
SpringerBriefs in Mathematics, https://doi.org/10.1007/978-3-032-05511-8_4

$$x(t) = E_\alpha(At^\alpha)x_0 + \frac{1}{\Gamma(\alpha)} \int_0^t \frac{1}{(t-s)^{1-\alpha}} E_{\alpha,\alpha}(A(t-s)^\alpha)h(s)\,ds, \ \forall t \geq 0.$$

$$(4.3)$$

This can be shown by using Laplace transforms, see e.g., Diethelm [1, Chap. 7] and the references therein.

Note that $x(t) = E_\alpha(\lambda t^\alpha)u$ is a solution of the linear Caputo FDE (4.1) with the initial condition $x(0) = u \in \mathbb{R}^d$ if and only if λ is an eigenvalue of the matrix A with the corresponding eigenvector u, since substituting into (4.1) leads to

$${}^C D_{0+}^\alpha E_\alpha(\lambda t^\alpha)u = A E_\alpha(\lambda t^\alpha)u = E_\alpha(\lambda t^\alpha)Au$$

while

$${}^C D_{0+}^\alpha E_\alpha(\lambda t^\alpha) = \lambda E_\alpha(\lambda t^\alpha) \implies {}^C D_{0+}^\alpha E_\alpha(\lambda t^\alpha)u = \lambda u E_\alpha(\lambda t^\alpha),$$

so $E_\alpha(\lambda t^\alpha)Au = \lambda u E_\alpha(\lambda t^\alpha)$. Cancelling $E_\alpha(\lambda t^\alpha) \neq 0$ from both sides then implies $Au = \lambda A$.

Theorem 4.1 *Suppose that the $d \times d$ matrix A has d linearly independent eigenvectors $u_1, \ldots, u_d$ with corresponding eigenvalues $\lambda_1, \ldots, \lambda_d$. Then, the solution (4.2) of the linear Caputo FDE (4.1) with the initial condition $x(0) = x_0 \in \mathbb{R}^d$ is*

$$x(t) = E_\alpha(At^\alpha)x_0 = \sum_{j=1^d} c_j u_j E_\alpha(\lambda_j t^\alpha),$$

where the constants $(c_1, \ldots, c_d)^T = [u_1|\cdots|u_d]^{-1}x_0$.

This is essentially Theorem 7.13 in Diethelm [1]. It corresponds to the case in which the matrix A is diagonalisable. Note that the eigenvalues need not be distinct and may be complex-valued (in which case they occur in complex-conjugate pairs).

More generally, the geometric multiplicity of an eigenvalue can be less than its algebraic multiplicity, i.e., the number of linearly independent eigenvectors is less than the multiplicity of the corresponding eigenvalue in the characteristic polynomial. Then, the Jordan canonical form of the matrix A needs to be used. This involves Jordan blocks of the form

$$J_k = \begin{pmatrix} \lambda_k & 1 & 0 & \cdots & 0 \\ 0 & \lambda_k & 1 & \cdots & 0 \\ \vdots & \vdots & \ddots & \ddots & \vdots \\ 0 & 0 & \cdots & \lambda_k & 1 \\ 0 & 0 & \cdots & 0 & \lambda_k \end{pmatrix}_{d_k \times d_k},$$

where λ_k is an eigenvalue. The component linear system

$${}^C D_{0+}^\alpha z_k(t) = J_k z_k(t), \ t > 0,$$

has the solution $z_k(t) = E_\alpha(J_k t^\alpha) z_k(0)$, where

$$E_\alpha(J_k t^\alpha) = \sum_{i=0}^{\infty} \frac{(J_k t^\alpha)^i}{\Gamma(\alpha i + 1)} = \sum_{i=0}^{\infty} \frac{t^{\alpha i} J_k^i}{\Gamma(\alpha i + 1)}$$

$$= \sum_{i=0}^{\infty} \frac{t^{\alpha i}}{\Gamma(i\alpha + 1)}
\begin{pmatrix}
C_i^0 \lambda_k^i & C_i^1 \lambda_k^{i-1} & C_i^2 \lambda_k^{i-2} & \cdots & C_i^{d_k-1} \lambda_k^{i-d_k+1} \\
0 & C_i^0 \lambda_k^i & C_i^1 \lambda_k^{i-1} & \cdots & C_i^{d_k-2} \lambda_k^{i-d_k+2} \\
\vdots & \cdots & \ddots & \cdots & \vdots \\
0 & 0 & \cdots & C_i^0 \lambda_k^i & C_i^1 \lambda_k^{i-1} \\
0 & 0 & \cdots & 0 & C_k^0 \lambda_i^k
\end{pmatrix}_{d_k \times d_k}$$

$$= \sum_{i=0}^{\infty} \frac{t^{\alpha i}}{\Gamma(i\alpha + 1)}
\begin{pmatrix}
\lambda_k^i & \dfrac{1}{1!}\dfrac{\partial \lambda_k^i}{\partial \lambda_k} & \dfrac{1}{2!}\dfrac{\partial^2 \lambda_k^i}{\partial \lambda_k^2} & \cdots & \dfrac{1}{(d_k-1)!}\dfrac{\partial^{d_k-1} \lambda_k^i}{\partial \lambda_k^{d_k-1}} \\
0 & \lambda_k^i & \dfrac{1}{1!}\dfrac{\partial \lambda_k^i}{\partial \lambda_k} & \cdots & \dfrac{1}{(d_k-2)!}\dfrac{\partial^{d_k-2} \lambda_k^i}{\partial \lambda_k^{d_k-2}} \\
\vdots & \cdots & \ddots & \cdots & \vdots \\
0 & 0 & \cdots & \lambda_k^i & \dfrac{1}{1!}\dfrac{\partial \lambda_k^i}{\partial \lambda_k} \\
0 & 0 & \cdots & 0 & \lambda_k^i
\end{pmatrix}_{d_k \times d_k},$$

here

$$C_i^j = \begin{cases} \dfrac{i!}{j!(i-j)!}, & \text{if } i \geq j, \\[2mm] 0, & \text{otherwise,} \end{cases}$$

which are the binomial coefficients for $j \leq i$. This implies

$$E_\alpha(J_k t^\alpha) =
\begin{pmatrix}
E_\alpha(\lambda_k t^\alpha) & \dfrac{1}{1!}\dfrac{\partial}{\partial \lambda_k} E_\alpha(\lambda_k t^\alpha) & \cdots & \dfrac{1}{(d_k-1)!}\dfrac{\partial^{d_k-1}}{\partial \lambda_k^{d_k-1}} E_\alpha(\lambda_k t^\alpha) \\
0 & E_{\alpha,\beta}(\lambda_i t^\alpha) & \cdots & \dfrac{1}{(d_k-2)!}\dfrac{\partial^{d_k-2}}{\partial \lambda_k^{d_k-2}} E_\alpha(\lambda_k t^\alpha) \\
\vdots & \cdots & \ddots & \vdots \\
0 & 0 & \cdots & E_\alpha(\lambda_k t^\alpha)
\end{pmatrix}_{d_k \times d_k},$$

cf. Lancaster and Tismenetsky [2, Theorem 4, p. 311]. From this it follows that

$$
E_\alpha(J_k t^\alpha) = \begin{pmatrix}
E_\alpha(\lambda_k t^\alpha) & t^\alpha E_\alpha^{(1)}(\lambda_k t^\alpha) & t^{2\alpha} E_\alpha^{(2)}(\lambda_k t^\alpha) & \cdots & t^{(d_k-1)\alpha} E_\alpha^{(d_k-1)}(\lambda_k t^\alpha) \\
0 & E_\alpha(\lambda_k t^\alpha) & t^\alpha E_\alpha^{(1)}(\lambda_k t^\alpha) & \cdots & t^{(d_k-2)\alpha} E_\alpha^{(d_k-2)}(\lambda_k t^\alpha) \\
\vdots & \vdots & \ddots & \ddots & \vdots \\
0 & 0 & \cdots & E_\alpha(\lambda_k t^\alpha) & t^\alpha E_\alpha^{(1)}(\lambda_k t^\alpha) \\
0 & 0 & \cdots & 0 & E_\alpha(\lambda_k t^\alpha)
\end{pmatrix}_{d_k \times d_k},
$$

here $E_\alpha^{(j)}(\cdot)$ denotes the classical jth derivative of the Mittag-Leffler function $E_\alpha(\cdot)$.

The solutions of the linear Caputo FDE (4.1) are linear combinations of such terms in the $E_\alpha(J_k t^\alpha)$, which form blocks on a block diagonal of the matrix $E_\alpha(At^\alpha)$. See Diethelm [1, Chap. 7] for more information and examples.

The concepts of stability and asymptotical stability for linear FDEs, e.g., [1, Definition 7.2, p. 157], are similar to those for ODEs.

Definition 4.1 The zero solution of the linear Caputo FDE (4.1) is called *stable* if for any $\varepsilon > 0$ there exists $\delta = \delta(\varepsilon) > 0$ such that $\|\Phi(t, x_0)\| < \varepsilon$ for all $t \geq 0$ for all $\|x_0\| < \delta$. It is called *unstable* if it is not stable.

Definition 4.2 The zero solution of the linear Caputo FDE (4.1) is called *asymptotically stable* if it is stable and there exists some $\delta_0 > 0$ such that $\lim_{t \to \infty} \|\Phi(t, x_0)\| = 0$ whenever $\|x_0\| < \delta_0$.

Remark 4.1 Due to the linear property of the system (4.1), it is easily shown that

(i) The trivial solution of (4.1) is stable if and only if all its solutions are bounded on the interval $[0, \infty)$.

(ii) The trivial solution of the system (4.1) is asymptotically stable if its all non-trivial solutions converge to the origin.

Combining the above results on the nature of the solutions with the asymptotic properties of Mittag-Leffler functions gives:

Theorem 4.2 [1, Theorem 7.20] *Consider the linear Caputo FDE (4.1) and let λ_1, ..., λ_d denote the eigenvalues of the $d \times d$ matrix A. Then, the zero solution is*

(i) *asymptotically stable if and only if all of eigenvalues of A satisfy $|\arg\lambda_j| > \alpha\pi/2$ for $j = 1, \ldots, d$;*

(ii) *stable if and only if all of eigenvalues of A satisfy $|\arg\lambda_j| \geq \alpha\pi/2$ for $j = 1, \ldots, d$ and all eigenvalues with $|\arg\lambda_j| = \alpha\pi/2$ have geometric multiplicity equal to their algebraic multiplicity.*

4.1   Lyapunov Exponents

This monograph focuses on "autonomous" Caputo FDEs, but here a linear Caputo FDE with a time-dependent matrix will be considered,

$$^C D_{0+}^{\alpha} x(t) = A(t)x(t), \ t > 0, \tag{4.4}$$

with the initial value $x(0) = x_0 \in \mathbb{R}^d$, where $A : \mathbb{R}_{\geq 0} \to \mathbb{R}^{d \times d}$ is continuous.

Such equations arise when linearising about a time dependent solution $x^*(t)$ of autonomous nonlinear Caputo FDEs

$$^C D_{0+}^{\alpha} x(t) = g(x(t)), \ t > 0$$

to investigate its stability. Then, $A(t) := \nabla g(x^*(t))$.

Assumption 4.1 $A : \mathbb{R}_{\geq 0} \to \mathbb{R}^{d \times d}$ is continuous and uniformly bounded, i.e.,

$$M := \sup_{t \in \mathbb{R}_{\geq 0}} \|A(t)\| < \infty. \tag{4.5}$$

Let $\Phi(\cdot, x_0)$ be the solution of the linear Caputo FDE (4.4).

Lemma 4.1 *Suppose that Assumption 4.1 holds. Then for any $x_0 \in \mathbb{R}^d$,*

$$\|\Phi(t, x_0)\| \leq E_{\alpha}(Mt^{\alpha})\|x_0\| \quad \text{for all } t \geq 0.$$

Proof The solution $\Phi(\cdot, x_0)$ of the linear Caputo FDE (4.4) has the integral representation

$$\Phi(t, x_0) = x_0 + \frac{1}{\Gamma(\alpha)} \int_0^t (t - s)^{\alpha - 1} A(s)\Phi(s, x_0) \, ds, \ t \geq 0. \tag{4.6}$$

Since $\|A(t)\| \leq M$ for all $t \in \mathbb{R}_{\geq 0}$, it follows that

$$\|\Phi(t, x_0)\| \leq \|x_0\| + \frac{M}{\Gamma(\alpha)} \int_0^t \frac{1}{(t - s)^{1 - \alpha}} \|\Phi(s, x_0)\| \, ds \quad \text{for all. } t \geq 0.$$

Then the fractional Gronwall inequality, Lemma 2.4, gives

$$\|\Phi(t, x_0)\| \leq E_{\alpha}(Mt^{\alpha})\|x_0\| \quad \text{for all } t \geq 0. \qquad \square$$

4.1.1 Classical Lyapunov Exponents

The classical Lyapunov exponent of a vector valued function $f : \mathbb{R}_{\geq 0} \to \mathbb{R}^d$ is defined by

$$\chi(f) := \limsup_{t \to \infty} \frac{1}{t} \log \|f(t)\|.$$

It is very useful in investigating the dynamics of solutions of linear ODEs, but is not so useful for Caputo FDEs (4.4) as it is always nonnegative.

Proposition 4.1 *Suppose that the linear Caputo FDE* (4.4) *satisfies Assumption 4.1. Then,*

$$\chi(\Phi(\cdot, x_0)) \geq 0 \quad \text{for all } x_0 \in \mathbb{R}^d \setminus \{0\}.$$

Proof Suppose a contrary that there exists $x_0 \in \mathbb{R}^d \setminus \{0\}$ such that

$$\lambda := \chi(\Phi(\cdot, x_0)) = \limsup_{t \to \infty} \frac{1}{t} \log \|\Phi(t, x_0)\| < 0. \tag{4.7}$$

Thus, there exists $K > 0$ and $T > 0$ such that

$$\|\Phi(t, x_0)\| < K e^{\frac{\lambda}{2}t} \qquad \text{for all } t \geq T. \tag{4.8}$$

From (4.8) and the fact that $\sup_{t \in \mathbb{R}_{\geq 0}} \|A(t)\| \leq M$, it follows that for $t > T$

$$\left| \int_0^t (t-s)^{\alpha-1} A(s) \Phi(s, x_0)\, ds \right| \leq M \sup_{s \in [0,T]} \|\Phi(s, x_0)\| \int_0^T (t-s)^{\alpha-1} ds$$
$$+ KM \int_0^t (t-s)^{\alpha-1} e^{\frac{\lambda}{2}s}\, ds.$$

Put $I_1(t) := \int_0^T (t-s)^{\alpha-1} ds$ and $I_2(t) := \int_0^t (t-s)^{\alpha-1} e^{\frac{\lambda}{2}s}\, ds$. It is easy to see

$$I_1(t) = \frac{1}{\alpha}(t^\alpha - (t-T)^\alpha) \to 0 \quad \text{as } t \to \infty.$$

On the other hand, a direct computation yields that

$$I_2(t) = \int_0^{t/2} (t-s)^{\alpha-1} e^{\frac{\lambda}{2}s}\, ds + \int_{t/2}^t (t-s)^{\alpha-1} e^{\frac{\lambda}{2}s}\, ds$$
$$\leq \left(\frac{2}{t}\right)^{1-\alpha} \int_0^{t/2} e^{\frac{\lambda}{2}s} ds + e^{\frac{\lambda}{4}t} \int_{t/2}^t (t-s)^{\alpha-1} ds$$
$$\leq \frac{2}{|\lambda|} \left(\frac{2}{t}\right)^{1-\alpha} + \frac{e^{\frac{\lambda}{4}t}}{\alpha} \left(\frac{t}{2}\right)^\alpha \to 0 \quad \text{as } t \to \infty.$$

Therefore,

$$\limsup_{t \to \infty} \int_0^t (t-s)^{\alpha-1} A(s) \Phi(s, x_0)\, ds = 0.$$

This with the integral representation (4.6) of $\Phi(\cdot, x_0)$ implies that $\limsup_{t \to \infty} \|\Phi(t, x_0)\| = \|x_0\|$, which contradicts (4.8). $\qquad\square$

Denote $d_j = \dim E_j$ and $d_i = \dim E_i - \dim E_{i+1}$ for $i = 1, \ldots, j-1$. The set $\widehat{\Sigma} := \{(\lambda_i, d_i), i = 1, \ldots, j\}$ is often called the *Lyapunov spectrum* of the linear Caputo FDE (4.4).

4.1.2 Fractional Lyapunov Exponents

The Mittag-Leffler functions $E_\alpha(\cdot)$ play an important role in investigating linear Caputo FDEs comparable to that of the exponential function for linear ODEs. It is thus natural to use them and their inverse function $\log_\alpha^M$ (see Lemma 1.3) to generalise a notion of a Lyapunov exponent to Caputo FDEs.

Definition 4.3 Let $f : \mathbb{R}_{\geq 0} \to \mathbb{R}^d$ be an arbitrary function. The *fractional Lyapunov exponent of order α of f* is defined as

$$\chi_\alpha(f) = \limsup_{t \to \infty} \frac{1}{t^\alpha} \log_\alpha^M \|f(t)\|. \tag{4.9}$$

Such fractional Lyapunov exponents can be more easily calculated in terms of classical elementary functions.

Theorem 4.3 *Let $f : \mathbb{R}_{\geq 0} \to \mathbb{R}^d$ be an arbitrary function. Then*

(i) $\chi_\alpha(f) > 0$ if and only if $\chi(f) > 0$, in which case

$$\chi_\alpha(f) = \chi(f)^\alpha = \limsup_{t \to \infty} \left(\frac{1}{t} \log \|f(t)\| \right)^\alpha. \tag{4.10}$$

(ii) $\chi_\alpha(f) < 0$ if and only if $\limsup_{t \to \infty} t^\alpha \|f(t)\| < \infty$, in which case

$$\chi_\alpha(f) = - \frac{1}{\Gamma(1 - \alpha) \limsup_{t \to \infty} t^\alpha \|f(t)\|}. \tag{4.11}$$

(iii) $\chi_\alpha(f) = 0$ if and only if

$$\chi(f) \leq 0 \quad \text{and} \quad \limsup_{t \to \infty} t^\alpha \|f(t)\| = \infty.$$

Proof (i) ($\Rightarrow$) Suppose that $\lambda := \chi_\alpha(f) > 0$. It will be shown that $\chi(f) > 0$ and that the equality (4.10) holds. Let $\varepsilon \in (0, \lambda)$ be arbitrary. Then, from Lemma 1.3(ii) and the definition of $\chi_\alpha(f)$,

$$\limsup_{t \to \infty} \frac{1}{t^\alpha} \log_\alpha^M (e^{(\lambda+\varepsilon)^{\frac{1}{\alpha}} t}) > \limsup_{t \to \infty} \frac{1}{t^\alpha} \log_\alpha^M (\|f(t)\|)$$

and

$$\limsup_{t \to \infty} \frac{1}{t^\alpha} \log_\alpha^M (\|f(t)\|) > \limsup_{t \to \infty} \frac{1}{t^\alpha} \log_\alpha^M (e^{(\lambda-\varepsilon)^{\frac{1}{\alpha}} t}),$$

which together with Lemma 1.3(i) implies that there exists $T > 0$ such that

$$e^{(\lambda+\varepsilon)^{\frac{1}{\alpha}}t} \geq \|f(t)\| \geq e^{(\lambda-\varepsilon)^{\frac{1}{\alpha}}t} \qquad \text{for all } t \geq T.$$

Consequently,

$$(\lambda + \varepsilon)^{\frac{1}{\alpha}} \geq \limsup_{t\to\infty} \frac{1}{t} \log \|f(t)\| \geq (\lambda - \varepsilon)^{\frac{1}{\alpha}}.$$

Letting $\varepsilon \to 0$ gives $\chi(f) = \chi_\alpha(f)^{\frac{1}{\alpha}}$, which proves (4.10).

(i) ($\Leftarrow$) Suppose that $\gamma := \chi(f) > 0$. It will be shown that $\chi_\alpha(f) > 0$. Indeed, by definition of $\chi(f)$, there exists $T > 0$ such that

$$\|f(t)\| \geq e^{\frac{\gamma}{2}t} \qquad \text{for all } t > T.$$

Together Lemma 1.3(i) and (ii) this implies that

$$\chi_\alpha(f) = \limsup_{t\to\infty} \frac{1}{t^\alpha} \log_\alpha^M(\|f(t)\|) \geq \limsup_{t\to\infty} \frac{1}{t^\alpha} \log_\alpha^M(e^{\frac{\gamma}{2}t}) \geq \frac{\gamma}{2} > 0,$$

which completes this part of the proof.

(ii) ($\Rightarrow$) Suppose that $\lambda := \chi_\alpha(f) < 0$. The aim now is to show that $\limsup_{t\to\infty} t^\alpha \|f(t)\| < \infty$ and that (4.11) holds. For this purpose, let $\varepsilon \in (0, -\lambda)$ be arbitrary. It follows from Lemma 1.3(iii) and definition of $\chi_\alpha(f)$ that

$$\limsup_{t\to\infty} \frac{1}{t^\alpha} \log_\alpha^M\left(\frac{1}{-(\lambda + \varepsilon)\Gamma(1 - \alpha)t^\alpha}\right) > \limsup_{t\to\infty} \frac{1}{t^\alpha} \log_\alpha^M(\|f(t)\|)$$

and

$$\limsup_{t\to\infty} \frac{1}{t^\alpha} \log_\alpha^M(\|f(t)\|) > \limsup_{t\to\infty} \frac{1}{t^\alpha} \log_\alpha^M\left(\frac{1}{-(\lambda - \varepsilon)\Gamma(1 - \alpha)t^\alpha}\right).$$

This together with Lemma 1.3(i) means that there exists $T > 0$ such that

$$\frac{1}{-(\lambda + \varepsilon)\Gamma(1 - \alpha)t^\alpha} \geq \|f(t)\| \geq \frac{1}{-(\lambda - \varepsilon)\Gamma(1 - \alpha)t^\alpha} \qquad \text{for } t \geq T.$$

Consequently,

$$\frac{1}{-(\lambda + \varepsilon)\Gamma(1 - \alpha)} \geq \limsup_{t\to\infty} t^\alpha \|f(t)\| \geq \frac{1}{-(\lambda - \varepsilon)\Gamma(1 - \alpha)}.$$

Letting $\varepsilon \to 0$ gives $\limsup_{t\to\infty} t^\alpha \|f(t)\| = \frac{1}{-\lambda\Gamma(1-\alpha)}$. This proves (4.11).

(ii) ($\Leftarrow$) Suppose that $\gamma := \limsup_{t\to\infty} t^\alpha \|f(t)\| < \infty$. It will be shown that $\chi_\alpha(f) < 0$. Indeed, let $T > 0$ such that

$$\|f(t)\| \le \frac{2\gamma}{t^\alpha} \qquad \text{for all } t > T.$$

This together Lemmas 1.3(i) and (iii) implies that

$$\limsup_{t\to\infty} \frac{1}{t^\alpha} \log_\alpha^M (\|f(t)\|) \le \limsup_{t\to\infty} \frac{1}{t^\alpha} \log_\alpha^M (2\gamma t^{-\alpha}) = -\frac{1}{2\gamma\,\Gamma(1-\alpha)} < 0,$$

which proves that $\chi_\alpha(f) < 0$ and hence completes this part of the proof.

(iii) This follows directly from (i) and (ii). $\square$

Fractional Lyapunov exponents have another useful property which facilitates their calculation.

Lemma 4.2 [3, Lemma 3.3] *The following statements hold.*

(i) For an arbitrary function $f : \mathbb{R}_{\ge 0} \to \mathbb{R}^d$ and $c \in \mathbb{R} \setminus \{0\}$,

$$\chi_\alpha(c\,f) = \begin{cases} \chi_\alpha(f) & \text{if } \chi_\alpha(f) > 0, \\[2mm] \dfrac{\chi_\alpha(f)}{|c|} & \text{if } \chi_\alpha(f) \le 0. \end{cases}$$

(ii) Let $f_1,\, f_2 : \mathbb{R}_{\ge 0} \to \mathbb{R}^d$ be arbitrary functions such that $\chi_\alpha(f_1) \ge 0$. Then,

$$\chi_\alpha(f_1 + f_2) \le \max\{\chi_\alpha(f_1), \chi_\alpha(f_2)\}$$

with equality if $\chi_\alpha(f_1) \ne \chi_\alpha(f_2)$.

(iii) Let $f_1,\, f_2 : \mathbb{R}_{\ge 0} \to \mathbb{R}^d$ be arbitrary functions such that $\chi_\alpha(f_1),\, \chi_\alpha(f_2) < 0$. Then, $\chi_\alpha(f_1 + f_2) < 0$.

Using Lemma 4.2(i) and linearity of $\Phi(t, \cdot)$, it follows for an arbitrary $x_0 \in \mathbb{R}^d \setminus \{0\}$ that

$$\chi_\alpha(\Phi(\cdot, x_0)) = \begin{cases} \chi_\alpha(\Phi(\cdot, \frac{x_0}{\|x_0\|})) & \text{if } \chi_\alpha(\Phi(\cdot, \frac{x_0}{\|x_0\|})) > 0, \\[3mm] \dfrac{1}{\|x_0\|} \chi_\alpha(\Phi(\cdot, \frac{x_0}{\|x_0\|})) & \text{if } \chi_\alpha(\Phi(\cdot, \frac{x_0}{\|x_0\|})) \le 0. \end{cases} \qquad (4.12)$$

Thus, in order to estimate the fractional Lyapunov exponent of solutions of the linear FDE(4.4), it suffices to estimate $\chi_\alpha(\Phi(\cdot, x_0))$ for $x_0 \in \mathbb{R}^d$ with $\|x_0\| = 1$. Lower and upper bounds on $\chi_\alpha(\Phi(\cdot, x_0))$ for $x_0 \in \mathbb{R}^d$ with $\|x_0\| = 1$ are provided in the following lemma.

Lemma 4.3 *Let $\mathbb{S}^{d-1} = \{x \in \mathbb{R}^d : \|x\| = 1\}$ denote the unit sphere in $\mathbb{R}^d$ and suppose that Assumption 4.1 holds. Then,*

$$\chi_\alpha(\Phi(\cdot, x_0)) \in [-M, M] \quad \text{for all } x_0 \in \mathbb{S}^{d-1},$$

where M is the bound in Assumption 4.1.

Proof Let $x_0 \in \mathbb{S}^{d-1}$ be arbitrary. It will first be shown that $\chi_\alpha(\Phi(\cdot, x_0)) \leq M$. Indeed, by virtue of Lemma 4.1,

$$\|\Phi(t, x_0)\| \leq E_\alpha(Mt^\alpha)\|x_0\| \quad \text{for all } t \geq 0.$$

Therefore,

$$\chi_\alpha(\Phi(., x_0)) = \limsup_{t \to \infty} \frac{1}{t^\alpha} \log_\alpha \|\Phi(t, x_0)\| \leq \limsup_{t \to \infty} \frac{1}{t^\alpha} \log_\alpha (E_\alpha(Mt^\alpha)\|x_0\|) \leq M.$$

To complete the proof, it suffices to show that $\chi_\alpha(\Phi(\cdot, x_0)) \geq -M$. Suppose the contrary, i.e., $\chi_\alpha(\Phi(\cdot, x_0)) \leq -M - 2\varepsilon$ for some $\varepsilon > 0$. By definition of χ_α, there exists $T > 0$ such that

$$\|\Phi(t, x_0)\| \leq E_\alpha(-(M + \varepsilon)t^\alpha) \quad \text{for all } t \geq T. \tag{4.13}$$

From the continuity of the function $\|\Phi(t, x_0)\|$ in $t \in [0, T]$, it follows that

$$K := \max_{t \in [0,T]} \|\Phi(t, x_0)\| < \infty. \tag{4.14}$$

Now (4.6) and the fact that $\|x_0\| = 1$ give

$$\|\Phi(t, x_0)\| + \frac{1}{\Gamma(\alpha)} \int_0^t \frac{1}{(t-s)^{1-\alpha}} \|A(s)\| \|\Phi(s, x_0)\| \, ds \geq 1,$$

which together with (4.13), (4.14) and the fact that $\sup_{t \geq 0} \|A(t)\| \leq M$ implies that

$$\limsup_{t \to \infty} \frac{1}{\Gamma(\alpha)} \left(\int_0^T \frac{KM}{(t-s)^{1-\alpha}} \, ds + \int_T^t \frac{M E_\alpha(-(M+\varepsilon)s^\alpha)}{(t-s)^{1-\alpha}} \, ds \right) \geq 1.$$

It follows from $\limsup_{t \to \infty} \int_0^T \frac{KM}{(t-s)^{1-\alpha}} \, ds = 0$ that

$$\limsup_{t \to \infty} \frac{1}{\Gamma(\alpha)} \int_0^t \frac{M E_\alpha(-(M+\varepsilon)s^\alpha)}{(t-s)^{1-\alpha}} \, ds \geq 1. \tag{4.15}$$

On the other hand, since $E_\alpha(-(M+\varepsilon)t^\alpha)$ is the solution of the linear Caputo FDE ${}^C D_{0+}^\alpha x(t) = -(M+\varepsilon)x(t)$ on $(0, \infty)$ with the initial condition $x(0) = 1$, it follows that

$$E_\alpha(-(M+\varepsilon)t^\alpha) + \frac{1}{\Gamma(\alpha)} \int_0^t \frac{(M+\varepsilon)E_\alpha(-(M+\varepsilon)s^\alpha)}{(t-s)^{1-\alpha}}\, ds = 1.$$

Hence,

$$\limsup_{t\to\infty} \frac{1}{\Gamma(\alpha)} \int_0^t \frac{M E_\alpha(-(M+\varepsilon)s^\alpha)}{(t-s)^{1-\alpha}}\, ds = \frac{M}{M+\varepsilon}.$$

This contradicts (4.15) and completes the proof. $\qquad\square$

4.1.3 Fractional Lyapunov Spectrum

The *fractional Lyapunov spectrum* of the linear Caputo FDE (4.4) is defined by

$$\Sigma_\alpha := \left\{ \chi_\alpha(\Phi(\cdot, x_0)) : x_0 \in \mathbb{R}^d \setminus \{0\} \right\}.$$

Theorem 4.4 *The fractional Lyapunov spectrum of the linear Caputo FDE (4.4) in $\mathbb{R}^d$ has the properties:*

(i) $\Sigma_\alpha \subset (-\infty, M]$.

(ii) The set $\Sigma_\alpha \cap \mathbb{R}_{\geq 0}$ consists of at most d distinct elements $\lambda_j < \lambda_{j-1} < \cdots < \lambda_1$.

(iii) If $\Sigma_\alpha \cap \mathbb{R}_{<0} \neq \emptyset$, then $\mathbb{R}_{<0} \subset \Sigma_\alpha$.

Moreover, the sets

$$S := \{x_0 \in \mathbb{R}^d : \chi_\alpha(\Phi(\cdot, x_0)) < 0\}, \tag{4.16}$$

and

$$E_i := \{x_0 \in \mathbb{R}^d : \chi_\alpha(\Phi(\cdot, x_0)) \leq \lambda_i\}, \qquad i = 1, \ldots, j,$$

are linear subspaces of $\mathbb{R}^d$ and there is a filtration $S =: E_{j+1} \subsetneqq E_j \subsetneqq E_{j-1} \subsetneqq \cdots \subsetneqq E_1$ such that

$$\chi_\alpha(\Phi(\cdot, x_0)) = \lambda_i \quad \text{if and only if} \quad x_0 \in E_i \setminus E_{i+1} \tag{4.17}$$

for $i = 1, \ldots, j$.

Proof (i) Let $x_0 \in \mathbb{R}^d \setminus \{0\}$ be arbitrary. Using (4.12) it follows that

$$
\chi_\alpha(\Phi(\cdot, x_0)) = \begin{cases} \chi_\alpha(\Phi(\cdot, \frac{x_0}{\|x_0\|})) & \text{if } \chi_\alpha(\Phi(\cdot, \frac{x_0}{\|x_0\|})) > 0, \\[2mm] \dfrac{\chi_\alpha(\Phi(\cdot, \frac{x_0}{\|x_0\|}))}{\|x_0\|} & \text{if } \chi_\alpha(\Phi(\cdot, \frac{x_0}{\|x_0\|})) \le 0, \end{cases}
$$

which together with Lemma 4.3 implies that $-\infty < \chi_\alpha(\Phi(\cdot, x_0)) \le M$. This completes the proof of this part.

(ii) Let $x, \tilde{x} \in \mathbb{R}^d \setminus \{0\}$ satisfy $\chi_\alpha(\Phi(\cdot, x)), \chi_\alpha(\Phi(\cdot, \tilde{x})) \ge 0$. By Lemma 4.2(i) and (ii),

$$
\chi_\alpha(\Phi(\cdot, x + \tilde{x})) \le \max\left\{\chi_\alpha(\Phi(\cdot, x)), \chi_\alpha(\Phi(\cdot, \tilde{x}))\right\}, \qquad \chi_\alpha(\Phi(\cdot, x)) = \chi_\alpha(\Phi(\cdot, c\,x))
$$

for all $c \in \mathbb{R} \setminus \{0\}$. This observation and the following argument shows that $\Sigma_\alpha \cap \mathbb{R}_{\ge 0}$ consists of at most d distinct elements.

Suppose the contrary that Σ_α consists of $d + 1$ distinct values $\lambda_{d+1} < \cdots < \lambda_1$. By the definition of Σ_α, there exist $x^1, \ldots, x^{d+1} \in \mathbb{R}^d \setminus \{0\}$ such that

$$
\chi(\Phi(\cdot, x^i)) = \lambda_i \qquad \text{for all } i = 1, \ldots, d + 1. \tag{4.18}
$$

Since $\dim(\mathbb{R}^d) = d$ it follows that $\sum_{j=1}^{d+1} \alpha_j x^j = 0$ for some $\alpha_1, \ldots, \alpha_{d+1} \in \mathbb{R}$ such that $\alpha_i \ne 0$ for some $i \in \{1, \ldots, d + 1\}$. Let $k := \min\{i \in \{1, \ldots, d + 1\} : \alpha_i \ne 0\}$. Thus, $x^k = \sum_{j=k+1}^{d+1} -\frac{\alpha_j}{\alpha_k} x^j$ and, by the linearity of the map $\Phi(t, \cdot)$,

$$
\Phi(t, x^k) = \sum_{j=k+1, \alpha_j \ne 0}^{d+1} -\frac{\alpha_j}{\alpha_k} \Phi(t, x^j).
$$

With the observation above this implies that

$$
\chi_\alpha(\Phi(\cdot, x^k)) \le \max\left\{\chi_\alpha\left(-\frac{\alpha_j}{\alpha_k}\Phi(\cdot, x^j)\right) : j \in \{k+1, \ldots, d+1\} \text{ with } \alpha_j \ne 0\right\}
$$
$$
= \max\left\{\chi_\alpha(\Phi(\cdot, x^j)) : j \in \{k+1, \ldots, d+1\} \text{ with } \alpha_j \ne 0\right\}.
$$

Therefore, by (4.18), $\lambda_k \le \max\{\lambda_j : j = k+1, \ldots, d+1\}$, which leads to a contradiction. This completes this part of the proof.

(iii) The fact that $\mathbb{R}_{<0} \subset \Sigma_\alpha$ provided $\Sigma_\alpha \cap \mathbb{R}_{<0} \ne \emptyset$ is straightforward by using Lemma 4.2(i).

It remains to show that the set S and E_i for $i = 1, \ldots, j$ are linear subspaces of $\mathbb{R}^d$. Clearly, $0 \in S$ and $\chi_\alpha(\Phi(\cdot, x)), \chi_\alpha(\Phi(\cdot, \tilde{x})) < 0$ for $x, \tilde{x} \in S \setminus \{0\}$. By Lemma 4.2(iii), $\chi_\alpha(\Phi(\cdot, x + \tilde{x})) < 0$ and therefore $x + \tilde{x} \in S$. Moreover, for any $c \in \mathbb{R} \setminus \{0\}$,

Lemma 4.2(i) gives $\chi_\alpha(\Phi(\cdot, cx)) < 0$ and thus $cx \in S$. Thus, S is a linear subspace of $\mathbb{R}^d$.

Similarly, using Lemmas 4.2(i) and (ii) it follows that E_i is a linear subspace of $\mathbb{R}^d$ for $i = 1, \ldots, j$.

Finally, the dynamical characterisation (4.17) follows directly from the definition of S and E_i (cf. Theorem [3, Theorem 3.1] for the classical Lyapunov exponent χ). This completes the proof of Theorem 4.4. $\qquad\square$

A finer structure of the fractional Lyapunov spectrum can be obtained by restricting to initial values on the unit sphere $\mathbb{S}^{d-1}$.

Theorem 4.5 [4, Theorem 3.2] *Let $\lambda_1, \ldots, \lambda_j$ and the subspace S be as in Theorem 4.4. Suppose that $S \neq \emptyset$. Then, the set Λ_α of all fractional Lyapunov exponents of solutions starting from the unit sphere has the form*

$$\Lambda_\alpha = [a, b] \cup \bigcup_{i=1}^{j} \{\lambda_i\},$$

where

$$a = \inf\{\chi_\alpha(\Phi(\cdot, x)) : x \in S \cap \mathbb{S}^{d-1}\}, \quad b = \sup\{\chi_\alpha(\Phi(\cdot, x)) : x \in S \cap \mathbb{S}^{d-1}\}.$$

4.1.4 Characterization of Stability

The fractional Lyapunov spectrum provides information about the stability and asymptotic stability of linear Caputo FDE (4.4). The Definitions 4.1 and 4.2 also apply for the time-dependent linear Caputo FDE (4.4).

Theorem 4.6 *Consider linear Caputo FDE (4.4). Then,*

(i) If the zero solution of the linear Caputo FDE (4.4) is stable, then $\Sigma_\alpha \subset (-\infty, 0]$.

(ii) If $\Sigma_\alpha \subset (-\infty, 0)$, then the zero solution of the linear Caputo FDE (4.4) is asymptotically stable.

Proof (i) Suppose that the zero solution of the linear Caputo FDE (4.4) is stable. Let $x_0 \in \mathbb{R}^d \setminus \{0\}$ be arbitrary. It needs to be shown that $\chi_\alpha(\Phi(\cdot, x_0)) \leq 0$. Suppose the opposite, i.e., $\chi_\alpha(\Phi(\cdot, x_0)) \geq 2\varepsilon$ for some $\varepsilon > 0$. By definition of the fractional Lyapunov exponent, there exists a sequence $(t_n)_{n\in\mathbb{N}}$ with $\lim_{n\to\infty} t_n = \infty$ such that

$$\|\Phi(t_n, x_0)\| \geq E_\alpha(\varepsilon t_n^\alpha) \quad \text{for all } n \in \mathbb{N}.$$

Consequently, given an arbitrary $\delta > 0$,

$$\|\Phi(t_n, \delta x_0/\|x_0\|)\| \geq \frac{\delta}{\|x_0\|} E_\alpha(\varepsilon t_n^\alpha) \quad \text{for all } n \in \mathbb{N},$$

which implies that $\limsup_{n\to\infty} \|\Phi(t_n, \delta x_0/\|x_0\|)\| = \infty$. Thus, the zero solution of the linear Caputo FDE (4.4) is not stable and the proof is complete.

(ii) Suppose that $\Sigma_\alpha \subset (-\infty, 0)$. Let $\|\cdot\|_2$ denote the standard Euclidean norm on $\mathbb{R}^d$. Since all norms on $\mathbb{R}^d$ are equivalent, there exists $L > 1$ such that

$$\frac{1}{L}\|x\|_2 \leq \|x\| \leq L\|x\|_2 \quad \text{for } x \in \mathbb{R}^d.$$

Let $(e_i)_{i=1,\ldots,d}$ denote the standard Euclidean basis of $\mathbb{R}^d$. For any $\delta > 0$, let $x_0 \in \mathbb{R}^d$ satisfies $\|x_0\| < \delta$. Thus $\|x_0\|_2 < L\delta$ and x_0 can be represented as

$$x_0 = \sum_{i=1}^d \gamma_i e_i \quad \text{with} \quad \sum_{i=1}^d \gamma_i^2 < L^2\delta^2.$$

By linearity of (4.4), $\Phi(t, x_0) = \sum_{i=1}^d \gamma_i \Phi(t, e_i)$. Therefore,

$$\|\Phi(t, x_0)\| < L\delta \sum_{i=1}^d \|\Phi(t, e_i)\| \quad \text{for all } t \in \mathbb{R}_{\geq 0}. \tag{4.19}$$

Since $\chi_\alpha(\Phi(\cdot, e_i)) < 0$ for $i = 1, \ldots, d$, it follows that $\lim_{t\to\infty} \Phi(t, e_i) = 0$. Thus, by (4.19), then $\lim_{t\to\infty} \Phi(t, x_0) = 0$.

It remains to show that zero solution of the linear Caputo FDE (4.4) is stable. Indeed, it follows from the continuity of $t \mapsto \Phi(t, e_i)$ and $\lim_{t\to\infty} \Phi(t, e_i) = 0$ that

$$K := \sup_{t\geq 0, i=1,\ldots,d} \|\Phi(t, e_i)\| < \infty,$$

which together with (4.19) implies that

$$\|\Phi(t, x_0)\| < KLd\delta \quad \text{for all } \|x_0\| < \delta.$$

Thus, the zero solution of the linear Caputo FDE (4.4) is stable. This completes the proof is Theorem 4.6. $\qquad\square$

4.2   Endnotes

See Diethelm [1, Chap. 7] for linear Caputo FDEs. The Lyapunov exponent results are based on based on the papers [3] and [4]. See also [5–7].

References

1. Diethelm, K.: The analysis of fractional differential equations, Springer Lecture Notes in Mathematics, vol. 2004. Springer, Heidelberg (2010)
2. Lancaster, P., Tismenetsky, M.: The theory of matrices. Second Edition: With Applications (Computer Science and Scientific Computing). Academic Press, San Diego (1985)
3. Cong, N.D., Doan, T.S., Tuan, H.T.: On fractional Lyapunov exponent for solutions of linear fractional differential equations. Fract. Calculus Appl. Anal. **17**(2), 285–306 (2014)
4. Cong, N.D., Doan, T.S., Tuan, H.T., Siegmund, S.: Structure of the fractional Lyapunov spectrum for linear fractional differential equations. Adv. Dyn. Syst. Appl. **9**, 149–159 (2014)
5. Bonilla, B., Rivero, M., Trujillo, J.J.: On systems of linear fractional differential equations with constant coefficients. Appl. Math. Comput. **187**, 68–78 (2007)
6. Cong, N.D., Tuan, H.T.: Generation of nonlocal dynamical systems by fractional differential equations. J. Integral Equ. Appl. **29**, 585–608 (2017). https://doi.org/10.1216/JIE-2017-29-4-585
7. Cong, N.D., Doan, T.S., Tuan, H.T.: An instability theorem of nonlinear fractional differential systems. Discr. Continuous Dyn. Syst. Ser. B **22**, 3079–3090 (2017)

Chapter 5
Asymptotic Stability and Instability

Abstract The stability and instability of equilibrium points of nonlinear autonomous fractional differential systems in finite-dimensional Euclidean spaces are investigated. The analysis is carried out by means of the linearization method. A distinctive feature compared with the classical case of ordinary differential equations is the construction of a Lyapunov–Perron operator together with a weighted norm adapted to the fractional framework. In addition to establishing the local convergence of solutions to the equilibrium, the rate of this convergence is also determined.

Keywords Nonlinear Caputo fractional differential equations · Linearization method · Lyapunov–Perron operator · Stability · Weighted norm · Banach fixed point theorem · Instability · Mittag-Leffler stability

The asymptotic properties of solutions of Caputo FDEs have some distinct features compared to those of solutions of ODEs. In particular, the solution of a Caputo FDE does not converge to an equilibrium point with exponential rate. This motivates the concept of Mittag-Leffler asymptotic stability for Caputo FDEs.

5.1 Solutions Cannot Decay Faster Than Power Rate

Consider a nonlinear Caputo FDE of order $\alpha \in (0, 1)$

$$^{C}D_{0+}^{\alpha}x(t) = g(x(t)) \tag{5.1}$$

in $\mathbb{R}^d$, where $g : \mathbb{R}^d \to \mathbb{R}^d$ satisfies the following assumptions.

Assumption 5.1 $g(0) = 0$.

Assumption 5.2 There exists $L > 0$ such that for all $x, y \in \mathbb{R}^d$,

$$\|g(x) - g(y)\| \le L\|x - y\|. \tag{5.2}$$

The initial value problem for the Caputo FDE (5.1) has unique solution in the whole $\mathbb{R}_{\geq 0}$ for any given initial value in $\mathbb{R}^d$ by Theorem 3.1. It will be shown that there is no nontrivial solution of (5.1) converging to the origin with exponential rate.

Lemma 5.1 *Every nontrivial solution of* (5.1) *does not converge to the origin with exponential rate.*

Proof Let $\varphi(\cdot, x_0)$ be the unique solution corresponding to an initial value $x_0 \neq 0$. Assume that this solution converges to the origin with the exponential rate, i.e., there exist positive constants λ and T_1 such that

$$\|\varphi(t, x_0)\| < e^{-\lambda t} \quad \text{for all } t \geq T_1. \tag{5.3}$$

Take and fix a positive number $K > 0$ satisfying

$$K\|x_0\| > 1. \tag{5.4}$$

From the asymptotic behaviour of the exponential functions and Mittag-Leffler functions, there is a constant $T_2 > 0$ such that

$$e^{-\lambda t} < \frac{1}{K}E_\alpha(-Lt^\alpha) \quad \text{for all } t \geq T_2. \tag{5.5}$$

Put $T_0 = \max\{T_1, T_2\}$. The equivalent integral equation representation of the Caputo FDE (5.1) is

$$\varphi(s, x_0) = x_0 + \frac{1}{\Gamma(\alpha)} \int_0^t (t - s)^{\alpha-1} g(\varphi(s, x_0)))ds \quad \text{for all } t \geq 0,$$

so by the global Lipschitz property, Assumption 5.2,

$$\|x_0\| - \|\varphi(t, x_0)\| \leq \|\varphi(t, x_0) - x_0\| \leq \frac{L}{\Gamma(\alpha)} \int_0^t (t - s)^{\alpha-1} \|\varphi(s, x_0)\| \, ds.$$

Hence using (5.3) and (5.5),

$$\frac{\Gamma(\alpha)\|x_0\|}{L} \leq \limsup_{t \to \infty} \int_0^{T_0} (t - s)^{\alpha-1} \|\varphi(s, x_0)\| \, ds$$

$$+ \limsup_{t \to \infty} \int_{T_0}^t (t - s)^{\alpha-1} \|\varphi(s, x_0)\| \, ds$$

$$\leq \sup_{s \in [0,T_0]} \|\varphi(s, x_0)\| \limsup_{t \to \infty} \int_0^{T_0} (t - s)^{\alpha-1} \, ds$$

$$+ \limsup_{t \to \infty} \int_{T_0}^t (t - s)^{\alpha-1} e^{-\lambda s} \, ds$$

$$\leq \sup_{s\in[0,T_0]} \|\varphi(s, x_0\| \limsup_{t\to\infty} \frac{t^\alpha - (t-T_0)^\alpha}{\alpha}$$

$$+ \limsup_{t\to\infty} \frac{1}{K} \int_0^t (t-s)^{\alpha-1} E_\alpha(-Ls^\alpha)\, ds$$

$$= \limsup_{t\to\infty} \frac{1}{K} \int_0^t (t-s)^{\alpha-1} E_\alpha(-Ls^\alpha)\, ds. \tag{5.6}$$

By (1.13)

$$E_\alpha(-Lt^\alpha) = 1 - \frac{L}{\Gamma(\alpha)} \int_0^t (t-s)^{\alpha-1} E_\alpha(-Ls^\alpha)\, ds,$$

so

$$\int_0^t (t-s)^{\alpha-1} E_\alpha(-Ls^\alpha)\, ds = \frac{\Gamma(\alpha)}{L}\left(1 - E_\alpha(-Lt^\alpha)\right) \to \frac{\Gamma(\alpha)}{L} \quad \text{as } t \to \infty.$$

This together (5.6) implies

$$K\|x_0\| \leq 1,$$

a contradiction to (5.4). Thus there do not exist any nontrivial solution of (5.1) converging to the origin with an exponential rate. The proof is complete. $\square$

Remark 5.1 Lemma 5.1 remains true if the global Lipschitz property, Assumption 5.2, is replaced by a weaker local Lipschitz property.

Assumption 5.3 There exists $L > 0$ such that for all $x, y \in \mathbb{R}^d$,

$$\|g(x) - g(y)\| \leq L_R \|x - y\| \quad \text{for all } x, y \in B_{\mathbb{R}^d}(0, R), \tag{5.7}$$

where $B_{\mathbb{R}^d}(0, R) := \{x \in \mathbb{R}^d : \|x\| \leq R\}$.

A closer look at the proof of Lemma 5.1 allows an even stronger statement on the decaying rate of solutions of fractional differential equations, namely a power rate of decay.

Theorem 5.1 *Any nontrivial solution of the Caputo FDE (5.1) cannot decay to 0 faster than $t^{-\alpha}$. More precisely, suppose that the initial value $x_0 \neq 0$ and let $\beta > 0$ be an arbitrary positive number satisfying $\beta > \alpha$. Then*

$$\limsup_{t\to+\infty} t^\beta \, \|\varphi(t, x_0)\| = +\infty.$$

Proof Assume to the contrary that there exists an $\beta > \alpha$ such that

$$\limsup_{t\to+\infty} t^\beta \, \|\varphi(t, x_0)\| = M < \infty.$$

Then it suffices to argue as in the proof of Lemma 5.1, modifying the relations (5.3) and (5.5) by changing $e^{-\lambda t}$ there to $t^\beta/(M+1)$, to derive a contradiction.   □

5.2   Stability Concepts for Caputo FDEs

Consider again the nonlinear Caputo FDE (5.1) in $\mathbb{R}^d$, where g satisfies Assumptions 5.1 and 5.3.

Since g is local Lipschitz continuous, Theorems 3.3 and 3.4 imply the existence of a unique solution to the initial value problem (5.1), $x(0) = x_0$ for $x_0 \in B_{\mathbb{R}^d}(0, R)$.

Let $\varphi : I \times \mathbb{R}^d \to \mathbb{R}^d$ denote the solution of (5.1) with initial value $x(0) = x_0$ on the maximal interval of existence $I := [0, T_b(x_0))$ with $0 < T_b(x_0) \le \infty$.

Well known concepts of stability and asymptotic stability of the zero solution of classical ordinary differential equations can be applied to Caputo FDEs, see [1, Definition 7.2, p. 157].

Definition 5.1   The zero solution of the Caputo FDE (5.1) is called *stable* if for any $\varepsilon > 0$ there exists $\delta = \delta(\varepsilon) > 0$ such that $T_b(x_0) = \infty$ and $\|\varphi(t, x_0)\| < \varepsilon$ for all $t \ge 0$ and $x_0 \in B_{\mathbb{R}^d}(0, \delta)$. It is called *unstable* if it is not stable.

Definition 5.2   The zero solution of the Caputo FDE (5.1) is called *asymptotically stable* if it is stable and there exists some $\tilde{\delta} > 0$ such that $\lim_{t \to \infty} \|\varphi(t, x_0)\| = 0$ whenever $\|x_0\| < \tilde{\delta}$.

Lemma 5.1 shows that the non-trivial solutions of Caputo FDEs decay with at most a power rate, so the concept of exponential stability for ODEs is not valid for Caputo FDEs. Following Chap. 2 the linear Caputo FDE

$$\begin{cases} {}^C D_{0+}^\alpha x(t) & = -\gamma x(t), \ t > 0 \\ x(0) & = x_0 \end{cases}$$

has the explicit solution $E_\alpha(-\gamma t^\alpha)x_0 \sim O(t^{-\alpha})$ as $t \to \infty$.

This suggests the concept of Mittag-Leffler asymptotic stability for Caputo FDEs.

Definition 5.3   The equilibrium point $x^* = 0$ of of the Caputo FDE (5.1) is called *Mittag-Leffler asymptotically stable* if there exist positive constants β, δ and M such that

$$\sup_{t \ge 0} t^\beta \, \|\varphi(t, x_0)\| \le M \tag{5.8}$$

for all $\|x_0\| \le \delta$.

Remark 5.2   In view of Theorem 5.1, the parameter β in the Definition 5.3 must satisfy $\beta \le \alpha$.

Lemma 5.2   *The Mittag-Leffler asymptotic stability is strictly stronger than the asymptotic stability.*

Proof Consider the nonlinear scalar Caputo FDE of order $0 < \alpha < 1$:

$$\begin{cases} {}^{C}D_{0+}^{\alpha}x(t) = g(x(t)), \ t > 0, \\ x(0) = x_0, \end{cases} \tag{5.9}$$

where

$$g(x) := \begin{cases} -e^{-1/x}x, & \text{if } x > 0, \\ 0, & \text{if } x = 0, \\ -e^{1/x}x, & \text{if } x < 0. \end{cases} \tag{5.10}$$

It is clear that $g \in C^2$ and that the scalar Caputo FDE (5.9)–(5.10) has a unique solution $\varphi(t, x_0)$ for each $x_0 \in \mathbb{R}$, which exists globally on $\mathbb{R}_{\geq 0}$.

Fix some $x_0 > 0$. By Theorem 7.1, the solution $\varphi(t, x_0)$ cannot intersect the trivial solution, hence $\varphi(t, x_0) > 0$ for all $t \in \mathbb{R}_{\geq 0}$.

Now let $n \geq 2$ be an arbitrary integer. Put $f(x) := -(n-1)!x^n$ on a neighbourhood of 0 and extend it suitably to obtain $f(x) \leq g(x)$ on $(0, \infty)$. By Proposition 2.6, the solution $\varphi(t, x_0)$ of (5.9)–(5.10) is bounded by the solution of the Caputo FDE

$$ {}^{C}D_{0+}^{\alpha}y(t) = f(y(t)), \quad y(0) = x_0. \tag{5.11}$$

Using the construction of a sub-solution by Vergara and Zacher [2, pp. 332–334], it shows that the solution $y(t, x_0)$ of the IVP (5.11) has decay rate of $t^{-\alpha/n}$. Hence the function $\varphi(t, x_0)$, which is larger or equal to $y(t, x_0)$, cannot converge faster than $t^{-\alpha/n}$. Since n is arbitrary, $\varphi(t, x_0)$ cannot decay with power-rate. Thus, the zero solution of the Caputo FDE (5.9)–(5.10) is not Mittag-Leffler asymptotically stable.

On the other hand, from the fact that $f_{|(0,\infty)} \in C^2(0, \infty)$, it follows from Feng et al. [3, Theorem 3.3] that the solution $\varphi(t, x_0)$ of (5.9)–(5.10) is strictly decreasing for $t \geq 0$. Suppose that there exists a $\delta \in (0, 1)$ such that $\varphi(t, x_0) \geq \delta$ for all $t \geq 0$. Then,

$$ {}^{C}D_{0+}^{\alpha}x(t) \leq -e^{-1/\delta}x(t), \quad t > 0. $$

Then by the comparison principle, Lemma 2.6,

$$\varphi(t, x_0) \leq x_0 E_{\alpha}(-e^{-1/\delta}t) \to 0 \quad \text{as} \quad t \to \infty,$$

which leads to a contradiction. Consequently, $\varphi(t, x_0)$ converges to 0 as t tends to ∞. A similar assertion is also true for the solution starting from any $x_0 < 0$.

Finally, since $g \in C^2$ the Caputo FDE (5.9)–(5.10), with the initial condition $x_0 = 0$ has the unique solution $\varphi(t, 0) \equiv 0$. Hence, the zero solution of Caputo FDE (5.9)–(5.10) is asymptotically stable. $\qquad\square$

5.3 Linearised Stability for Nonlinear Caputo FDEs

Consider a nonlinear Caputo fractional differential equation

$$^{C}D_{0+}^{\alpha}x(t) = Ax(t) + g(x(t)) \tag{5.12}$$

in $\mathbb{R}^d$, where A is a $d \times d$ matrix and $g : \mathbb{R}^d \to \mathbb{R}^d$ is continuous on $\mathbb{R}^d$ and Lipschitz continuous in a neighborhood of the origin.

Assumption 5.4

$$g(0) = 0 \quad \text{and} \quad \lim_{R \to 0} L_g(R) = 0, \tag{5.13}$$

where

$$L_g(R) := \sup_{x,y \in B_{\mathbb{R}^d}(0,R)} \frac{\|g(x) - g(y)\|}{\|x - y\|}.$$

Let $\sigma(A) = \{\lambda_1, \dots, \lambda_n\}$ denote the spectrum of a matrix A, where $\lambda_1, \dots, \lambda_n$ are its eigenvalues.

Assumption 5.5

$$\sigma(A) \subset \Lambda_{\alpha}^{s} := \left\{ \lambda \in \mathbb{C} \setminus \{0\} : |\arg(\lambda)| > \frac{\alpha\pi}{2} \right\}. \tag{5.14}$$

Under Assumption 5.5 the zero solution of the linear Caputo FDE

$$^{C}D_{0+}^{\alpha}z(t) = Az(t)$$

is Mittag-Leffler asymptotically stable [1, Theorem 7.2].

The Mittag-Leffler asymptotic stability of the zero solution of a nonlinear Caputo FDE is determined by that of its the linear part.

Theorem 5.2 *Assume that A satisfies (5.14) in Assumption 5.5 and that g satisfies Assumption 5.4. Then the zero solution of the Caputo FDE (5.12) is Mittag-Leffler asymptotically stable.*

The proof of Theorem 5.2, given in the next subsection, requires several technical lemmas on the properties of the Mittag-Leffler functions.

Lemma 5.3 *(i) For any $\lambda \in \Lambda_{\alpha}^{s}$, there exists a constant $C_1 > 0$ such that*

$$|E_{\alpha}(\lambda t^{\alpha})| \leq C_1 E_{\alpha}(-t^{\alpha}), \quad t \geq 0.$$

(ii) There exists a constant $C_2 > 0$ such that

$$t^{\alpha} \int_0^t (t - s)^{\alpha-1} E_{\alpha,\alpha}(-(t - s)^{\alpha}) s^{-\alpha} \, ds \leq C_2, \quad t \geq 0.$$

Proof The proof of (i) follows from Theorem 1.4 (i) and the fact that

$$\lim_{t \to \infty} \frac{|E_\alpha(\lambda t^\alpha)|}{E_\alpha(-t^\alpha)} = C,$$

a constant, while (ii) follows from the identity (1.13) and the asymptotic behaviour of Mittag-Leffler function $E_\alpha(-t^\alpha)$. $\qquad\square$

Lemma 5.4 [4, Theorem 3(ii)] *Let $\lambda \in \Lambda_\alpha^s$. Then, there exists a positive constant C_3 such that*

$$\int_0^\infty s^{\alpha-1} |E_{\alpha,\alpha}(\lambda s^\alpha)|\, ds < C_3.$$

Proof of Theorem 5.2

Proof The proof consists of three main steps.
Step 1: Transformation of the linear part

Here the matrix A in (5.12) is transformed to its Jordan normal form. For simplicity, $\mathbb{R}$ is embedded into $\mathbb{C}$ and A is considered as a complex-valued matrix leading to its Jordan form below before returning to $\mathbb{R}$ (this technique is well known in the theory of ordinary differential equations, where further discussion can be found).

By linear algebra there is a nonsingular matrix $T \in \mathbb{C}^{d \times d}$ such that

$$T^{-1}AT = \operatorname{diag}(J_1, \ldots, J_n),$$

where for $i = 1, \ldots, n$ the Jordan block J_i has the form

$$J_i = \lambda_i \operatorname{id}_{d_i \times d_i} + \delta_i N_{d_i \times d_i},$$

with λ_i is an eigenvalue, $\delta_i \in \{0, 1\}$ and the nilpotent matrix $N_{d_i \times d_i}$ is given by

$$N_{d_i \times d_i} := \begin{pmatrix} 0 & 1 & 0 & \cdots & 0 \\ 0 & 0 & 1 & \cdots & 0 \\ \vdots & \vdots & \ddots & \ddots & \vdots \\ 0 & 0 & \cdots & 0 & 1 \\ 0 & 0 & \cdots & 0 & 0 \end{pmatrix}_{d_i \times d_i}. \tag{5.15}$$

Let η be an arbitrary, but fixed positive number. Applying the transformation $P_i := \operatorname{diag}(1, \eta, \ldots, \eta^{d_i-1})$ leads to

$$P_i^{-1} J_i P_i = \lambda_i \operatorname{id}_{d_i \times d_i} + \eta_i N_{d_i \times d_i}$$

wirh $\eta_i \in \{0, \eta\}$.

Hence, under the coordinate transformation $y := (TP)^{-1}x$ the system (5.12) becomes

$$^C D_{0+}^\alpha y(t) = \mathrm{diag}(\Lambda_1, \ldots, \Lambda_n) y(t) + h(y(t)), \tag{5.16}$$

where $\Lambda_i := \lambda_i \mathrm{id}_{d_i \times d_i}$ for $i = 1, \ldots, n$ and the function h is given by

$$h(y) := \mathrm{diag}(\eta_1 N_{d_1 \times d_1}, \ldots, \eta_n N_{d_n \times d_n}) y + (TP)^{-1} f(TPy).$$

Note that the mapping

$$x \mapsto \mathrm{diag}(\eta_1 N_{d_1 \times d_1}, \ldots, \eta_n N_{d_n \times d_n}) x$$

is a Lipschitz continuous function with the Lipschitz constant η. Thus, by (5.13)

$$h(0) = 0, \qquad \lim_{R \to 0} L_h(R) = \begin{cases} \eta & \text{if there exists } \eta_i = \eta, \\ 0 & \text{otherwise.} \end{cases} \tag{5.17}$$

Note also that the zero solutions of the Caputo FDEs equations (5.12) and (5.16) have the same stability and instability properties.

Step 2: Construction of a Lyapunov–Perron operator

Focus now on the transformed Caputo FDE (5.16). For any $x = (x^1, \ldots, x^n) \in \mathbb{C}^d = \mathbb{C}^{d_1} \times \cdots \times \mathbb{C}^{d_n}$, define the operator

$$\mathcal{T}_x : C([0, \infty), \mathbb{C}^d) \to C([0, \infty), \mathbb{C}^d)$$

by

$$(\mathcal{T}_x \xi)(t) = ((\mathcal{T}_x \xi)^1(t), \ldots, (\mathcal{T}_x \xi)^n(t)), \quad t \geq 0.$$

where

$$(\mathcal{T}_x \xi)^i(t) = E_\alpha(t^\alpha J_i) x^i + \int_0^t (t - \tau)^{\alpha-1} E_{\alpha,\alpha}((t - \tau)^\alpha J_i) h^i(\xi(\tau)) \, d\tau,$$

for $i = 1, \ldots, n$. The operator $\mathcal{T}_x$ is called the *Lyapunov–Perron operator* associated with (5.16).

Any solution of the nonlinear system (5.12) and its transformed counterpart (5.16) can be interpreted as a fixed point of the operators $\mathcal{T}_x$. This follows by a matrix version of the nonlinear variation of constants formula in Lemma 3.5, see [1, Remark 7.1, p. 135].

Define the weighted norm $\| \cdot \|_w$ on $C([0, \infty), \mathbb{C}^d)$ by

$$\|x\|_w = \max\{ \sup_{t \in [0,1]} \|x(t)\|, \ \sup_{t \geq 1} t^\alpha \|x(t)\| \}.$$

Then $C_w := \left\{ x \in C([0, \infty), \mathbb{C}^d) : \|x\|_w < \infty \right\}$ is also a Banach space with the norm $\| \cdot \|_w$.

Some estimates of the operator $\mathcal{T}_x$ in the space C_w will be derived in the following proposition and lemma.

Proposition 5.1 *Consider the transformed Caputo FDE* (5.16) *and suppose that the eigenvalues* $\lambda_1, \ldots, \lambda_n$ *of the matrix A satisfy Assumption 5.5. Then, there exists a constant* $C(\alpha, A)$ *depending on* α *and the eigenvalues* $\lambda_1, \ldots, \lambda_n$ *such that*

$$\|\mathcal{T}_x \xi - \mathcal{T}_{\hat{x}} \hat{\xi}\|_w \leq \max_{1 \leq i \leq n} \left\{ \sup_{t \in [0,1]} |E_\alpha(\lambda_i t^\alpha)| + \sup_{t \geq 1} t^\alpha |E_\alpha(\lambda_i t^\alpha)| \right\} \|x - \hat{x}\|$$
$$+ C(\alpha, A)\, L_h \left(\max\{\|\xi\|_\infty, \|\hat{\xi}\|_\infty\} \right) \|\xi - \hat{\xi}\|_w \qquad (5.18)$$

for all $x, \hat{x} \in \mathbb{C}^d$ *and* $\xi, \hat{\xi} \in C_w$.

Consequently, $\mathcal{T}_x$ *is well-defined as an operator on the Banach space* C_w *endowed with the norm* $\| \cdot \|_w$ *and*

$$\|\mathcal{T}_x \xi - \mathcal{T}_x \hat{\xi}\|_w \leq C(\alpha, A)\, L_h \left(\max\{\|\xi\|_\infty, \|\hat{\xi}\|_\infty)\} \right) \|\xi - \hat{\xi}\|_w.$$

Proof For $i = 1, \ldots, n$,

$$\|(\mathcal{T}_x \xi)^i(t) - (\mathcal{T}_{\hat{x}} \hat{\xi})^i(t)\| \leq \|x - \hat{x}\| |E_\alpha(\lambda_i t^\alpha)|$$
$$+ L_h(\max\{\|\xi\|_\infty, \|\hat{\xi}\|_\infty\}) \int_0^t (t - \tau)^{\alpha-1} |E_{\alpha,\alpha}(\lambda_i (t - \tau)^\alpha)| \|(\xi - \hat{\xi})(\tau)\|\, d\tau.$$

In particular, for $t \in [0, 1]$,

$$\sup_{t \in [0,1]} \|(\mathcal{T}_x \xi - \mathcal{T}_x \hat{\xi})^i(t)\| \leq \sup_{t \in [0,1]} |E_\alpha(\lambda_i t^\alpha)| \|x - \hat{x}\|$$
$$+ L_h(\max\{\|\xi\|_\infty, \|\hat{\xi}\|_\infty\}) \int_0^\infty u^{\alpha-1} |E_{\alpha,\alpha}(-\lambda_i u^\alpha)|\, du\, \|\xi - \hat{\xi}\|_w.$$
$$(5.19)$$

Furthermore,

$$\sup_{t \geq 1} t^\alpha \|(\mathcal{T}_x \xi - \mathcal{T}_x \hat{\xi})^i(t)\| \leq \sup_{t \geq 1} t^\alpha |E_\alpha(\lambda_i t^\alpha)| \, \|x - \hat{x}\| + C_{\lambda_i} L_h(\max\{\|\xi\|_\infty, \|\hat{\xi}\|_\infty\}) \times$$
$$\times \sup_{t \geq 1} t^\alpha \int_0^t (t - \tau)^{\alpha-1} E_{\alpha,\alpha}(-(t - \tau)^\alpha) \tau^{-\alpha}\, d\tau \|\xi - \hat{\xi}\|_w,$$
$$(5.20)$$

where the constant C_{λ_i} is chosen as in Lemma 5.3 (i).

Combining Lemmas 5.3, 5.4, Eqs. (5.19) and (5.20) then gives

$$\|\mathcal{T}_x\xi) - \mathcal{T}_{\hat{x}}\hat{\xi}\|_w \leq \max_{1\leq i\leq n}\left\{\sup_{t\in[0,1]}|E_\alpha(\lambda_i t^\alpha)| + \sup_{t\geq 1}t^\alpha|E_\alpha(\lambda_i t^\alpha)|\right\}\|x - \hat{x}\|$$
$$+ C(\alpha, A)\, L_h\left(\max\{\|\xi\|_\infty, \|\hat{\xi}\|_\infty\}\right)\|\xi - \hat{\xi}\|_w,$$

where

$$C(\alpha, A) := \max_{1\leq i\leq n}\int_0^\infty u^{\alpha-1}|E_{\alpha,\alpha}(\lambda_i u^\alpha)|\, du$$
$$+ C_\lambda \sup_{t\geq 1} t^\alpha \int_0^t (t-\tau)^{\alpha-1} E_{\alpha,\alpha}(-(t-\tau)^\alpha)\tau^{-\alpha}\, d\tau$$

with $C_\lambda := \max\{C_{\lambda_1}, \ldots, C_{\lambda_n}\}$. This completes the proof of Proposition 5.1. $\square$

This shows that $\mathcal{T}_x(\cdot)$ is well-defined and Lipschitz continuous with the constant $C(\alpha, A)$. Moreover, $C(\alpha, A)$ is independent of the constant η. Henceforth, choose $\eta = \frac{1}{2C(\alpha,A)}$.

Lemma 5.5 *Suppose that the eigenvalues $\lambda_1, \ldots, \lambda_n$ of the matrix A satisfy Assumption 5.5 and let $C(\alpha, A)$ be the constant defined in Proposition 5.1. Then*

(i) There is an $R > 0$ such that

$$q := C(\alpha, A)\, L_h(R) < 1. \tag{5.21}$$

(ii) Choose $R > 0$ satisfying (5.21) and let

$$\gamma := \max_{1\leq i\leq n}\left\{\sup_{t\in[0,1]}|E_\alpha(\lambda_i t^\alpha)| + \sup_{t\geq 1}t^\alpha|E_\alpha(\lambda_i t^\alpha)|\right\}$$

and

$$R^* := \frac{R(1-q)}{\gamma}. \tag{5.22}$$

Define $B_{C_w}(0, R) := \{\xi \in C_\infty([0,\infty), \mathbb{C}^d) : \|\xi\|_w \leq R\}$. Then, $\mathcal{T}_x (B_{C_w}(0, R)) \subset B_{C_w}(0, R)$ and

$$\|\mathcal{T}_x\xi - \mathcal{T}_x\hat{\xi}\|_w \leq q\,\|\xi - \hat{\xi}\|_w, \quad \xi, \hat{\xi} \in B_{C_w}(0, R)$$

for any $x \in B_{\mathbb{C}^d}(0, R^*)$.

Proof By (5.17), $\lim_{R\to 0} L_h(R) \leq \eta$. Since $\eta C(\alpha, A) = \frac{1}{2}$, the statement (i) is verified.

Now let $x \in B_{\mathbb{C}^d}(0, R^*)$ and $\xi \in B_{C_w}(0, R)$. Then by (5.18) in Proposition 5.1,

$$
\begin{aligned}
\|\mathcal{T}_x \xi\|_w \;\; &\leq\; \max_{1 \leq i \leq n} \left\{ \sup_{t \in [0,1]} |E_\alpha(\lambda_i t^\alpha)| + \sup_{t \geq 1} t^\alpha |E_\alpha(\lambda_i t^\alpha)| \right\} \|x\| \\
&\quad + C(\alpha, A)\, L_h(R) \|\xi\|_w \\
&\leq\; (1-q)R + qR,
\end{aligned}
$$

which proves that $\mathcal{T}_x(B_{C_w}(0, R)) \subset B_{C_w}(0, R)$. Moreover, from Proposition 5.1 and part (i),

$$
\|\mathcal{T}_x \xi - \mathcal{T}_x \hat{\xi}\|_w \leq C(\alpha, A) L_h(R)\, \|\xi - \hat{\xi}\|_w \leq q \|\xi - \hat{\xi}\|_w
$$

for all $x \in B_{\mathbb{C}^d}(0, R^*)$ and $\xi, \hat{\xi} \in B_{C_w}(0, R)$. This completes the proof of Lemma 5.5. $\qquad\square$

Step 3: Finishing of the proof of Theorem 5.2

It suffices to prove the Mittag-Leffler asymptotic stability for the zero solution of the transformed Caputo FDE (5.16). Let R^* be the constant defined in (5.22). For any $x \in B_{\mathbb{C}^d}(0, R^*)$, by Lemma 5.5 and the Contraction Mapping Principle, there is a unique fixed point $\xi \in B_{C_w}(0, R)$ of $\mathcal{T}_x$. This fixed point is also the unique solution of (5.16) satisfying $\xi(0) = x$.

Together with the existence and uniqueness of solutions for initial value problems for the Caputo FDE (5.16) in a neighbourhood of the origin, this shows that the trivial solution is stable in the Lyapunov sense. Furthermore,

$$
\sup_{t \geq 0} t^\alpha \|\xi(t)\| \leq R, \quad t \geq 0,
$$

which shows that the zero solution of the Caputo FDE (5.16) is Mittag-Leffler asymptotically stable. This completes the proof of Theorem 5.2. $\qquad\square$

5.4 Instability

Consider a nonlinear Caputo fractional differential equation

$$
{}^C D_{0+}^\alpha x(t) = Ax(t) + g(x(t)), \tag{5.23}
$$

where A is a $d \times d$ matrix and $g : \mathbb{R}^d \to \mathbb{R}^d$ is continuous on $\mathbb{R}^d$ and Lipschitz continuous in a neighbourhood of the origin, specifically let g satisfy.

Assumption 5.6

$$g(0) = 0 \quad \text{and} \quad \lim_{R \to 0} L_g(R) = 0, \tag{5.24}$$

where

$$L_g(R) := \sup_{x,y \in B_{\mathbb{R}^d}(0,R)} \frac{\|g(x) - g(y)\|}{\|x - y\|}.$$

Let $\sigma(A) = \{\lambda_1, \ldots, \lambda_n\}$ denote the spectrum of a matrix A, where $\lambda_1, \ldots, \lambda_n$ are its distinct eigenvalues and define

$$\Lambda_\alpha^u := \left\{\lambda \in \mathbb{C} \setminus \{0\} : |\arg(\lambda)| < \frac{\alpha\pi}{2}\right\}. \tag{5.25}$$

It will be assumed now that the following assumption holds.

Assumption 5.7 $\sigma(A) \cap \Lambda_\alpha^u \neq \emptyset$.

Under Assumption 5.7 the zero solution of the linear Caputo FDE

$$^{C}D_{0+}^\alpha z(t) = Az(t)$$

is unstable [1, Theorem 7.2]. Moreover, the zero solution of the nonlinear Caputo FDE (5.23) is also unstable.

Theorem 5.3 *Consider the nonlinear Caputo FDE (5.23) in $\mathbb{R}^d$ and suppose that Assumptions 5.6 and 5.7 hold. Then, the zero solution of nonlinear Caputo FDE (5.23) is unstable.*

Several technical lemmas on the properties of the Mittag-Leffler functions will be required in the proof of Theorem 5.3 given in the following subsection.

Lemma 5.6 [5, Lemma 2] *Let $\lambda \in \Lambda_\alpha^u$. Then there exist a real number $t_0 > 0$ and a positive constant $M(\alpha, \lambda)$ such that*

$$\left| E_\alpha(\lambda t^\alpha) - \frac{1}{\alpha} e^{\lambda^{\frac{1}{\alpha}} \tau} \right| \leq \frac{M(\alpha, \lambda)}{t^\alpha},$$

$$\left| t^{\alpha-1} E_{\alpha,\alpha}(\lambda t^\alpha) - \frac{1}{\alpha} \lambda^{\frac{1}{\alpha}-1} e^{\lambda^{\frac{1}{\alpha}} \tau} \right| \leq \frac{M(\alpha, \lambda)}{t^{\alpha+1}}$$

for every $t \geq t_0$.

Lemma 5.7 [5, Lemma 3] *Let $\lambda \in \Lambda_\alpha^u$. Then there exists a positive constant $K(\alpha, \lambda)$ such that*

$$\int_t^\infty \left| \lambda^{\frac{1}{\alpha}-1} E_\alpha(\lambda t^\alpha) e^{-\lambda^{\frac{1}{\alpha}}\tau} g(\tau) \right| d\tau \le K(\alpha, \lambda)\|g\|_\infty,$$

$$\int_0^t \left| \left((t-\tau)^{\alpha-1} E_{\alpha,\alpha}(\lambda(t-\tau)^\alpha) - \lambda^{\frac{1}{\alpha}-1} E_\alpha(\lambda t^\alpha) e^{-\lambda^{\frac{1}{\alpha}}\tau} \right) g(\tau) \right| d\tau \le K(\alpha, \lambda)\|g\|_\infty$$

for all $t \ge 0$ and any function $g \in C_\infty([0, \infty); \mathbb{C})$. $\qquad\square$

Lemma 5.8 [5, Lemma 4] *For any function $g \in C_\infty([0, \infty); \mathbb{C})$ and $\lambda \in \Lambda_\alpha^u$,*

$$\lim_{t \to \infty} \int_0^t (t-\tau)^{\alpha-1} \frac{E_{\alpha,\alpha}(\lambda(t-\tau)^\alpha)}{E_\alpha(\lambda t^\alpha)} g(\tau) \, d\tau = \lambda^{\frac{1}{\alpha}-1} \int_0^\infty e^{-\lambda^{\frac{1}{\alpha}}\tau} g(\tau) \, d\tau. \quad (5.26)$$

Proof of Theorem 5.3

Proof The proof consists of three main steps corresponding to those in the proof of Theorem 5.2.

Step 1: Transformation of the linear part

Let $T \in \mathbb{C}^{d \times d}$ be a nonsingular matrix transforming A into its Jordan normal form, i.e.,

$$T^{-1}AT = \text{diag}(J_1, \dots, I_n),$$

where for $i = 1, \dots, n$ the Jordan block J_i is of the following form

$$J_i = \lambda_i \, \text{id}_{d_i \times d_i} + \delta_i \, N_{d_i \times d_i},$$

where $\delta_i \in \{0, 1\}$, $\lambda_i \in \sigma(A)$, and the nilpotent matrix $N_{d_i \times d_i}$ is given by (5.15).

Let γ be an arbitrary but fixed positive number. Then, using the transformation $P_i := \text{diag}(1, \gamma, \dots, \gamma^{d_i - 1})$,

$$P_i^{-1} J_i P_i = \lambda_i \, \text{id}_{d_i \times d_i} + \gamma_i \, N_{d_i \times d_i},$$

$\gamma_i \in \{0, \gamma\}$. Denote $P := \text{diag}(P_1, \dots, P_n)$. Then under the transformation $y := (TP)^{-1}x$ system (5.23) becomes

$$^C D_{0+}^\alpha y(t) = \text{diag}(\Lambda_1, \dots, \Lambda_n)y(t) + h(y(t)) = \Lambda y(t) + h(y(t)), \qquad (5.27)$$

where $\Lambda := \text{diag}(\Lambda_1, \dots, \Lambda_n)$, $\Lambda_i := \lambda_i \text{id}_{d_i \times d_i}$ for $i = 1, \dots, n$ and the function h is given by

$$h(y) := \mathrm{diag}(\gamma_1 N_{d_1 \times d_1}, \ldots, \gamma_n N_{d_n \times d_n})y + (TP)^{-1} f(TPy). \tag{5.28}$$

Note that

$$h(0) = 0, \qquad \lim_{R \to 0} L_h(R) = \begin{cases} \gamma & \text{if there exists } \gamma_i = \gamma, \\ 0 & \text{otherwise.} \end{cases} \tag{5.29}$$

By Assumption 5.7 there is at least one eigenvalue $\lambda_i \in \Lambda_\alpha^u$. Without loss of generality, it can be assumed that $\lambda_i \in \Lambda_\alpha^u$ for $i = 1, \ldots, k^*$, where $1 \le k^* \le m$. Consequently, for simplicity of notation, (5.27) can be rewritten in the form

$$^C D_{0+}^\alpha y(t) = \mathrm{diag}(\mu_1, \ldots, \mu_d)y(t) + h(y(t)), \tag{5.30}$$

where h is defined by (5.28) and

$$\begin{aligned} \mu_i &\in \sigma(A) = \{\lambda_1, \ldots, \lambda_m\}, & i &= 1, \ldots, d, \\ \mu_i &\in \Lambda_\alpha^u, & i &= 1, \ldots, k^*, \\ \mu_i &\in \sigma(A) \setminus \Lambda_\alpha^u, & i &= k^* + 1, \ldots, d. \end{aligned} \tag{5.31}$$

Step 2: Construction of a Lyapunov–Perron operator

Let $C_\infty([0, \infty); \mathbb{C}^d)$ be the linear space of all continuous functions $\xi : [0, \infty) \to \mathbb{C}^d$ such that

$$\|\xi\|_\infty := \sup_{t \in \mathbb{R}_{\ge 0}} \|\xi(t)\| < \infty.$$

Clearly, $(C_\infty([0, \infty); \mathbb{C}^d), \|\cdot\|_\infty)$ is a Banach space. For any $\xi \in C_\infty([0, \infty); \mathbb{C}^d)$, the Lyapunov–Perron operator $\mathcal{T} : C_\infty([0, \infty); \mathbb{C}^d) \to C([0, \infty); \mathbb{C}^d)$ be defined by

$$\mathcal{T}\xi(t) := ((\mathcal{T}\xi)^1(t), \ldots, (\mathcal{T}\xi)^d(t))^{\mathrm{T}} \quad \text{for all} \ \ t \ge 0, \tag{5.32}$$

where

$$\begin{aligned} (\mathcal{T}\xi)^i(t) := &\int_0^t (t - \tau)^{\alpha-1} E_{\alpha,\alpha}(\mu_i(t - \tau)^\alpha) h^i(\xi(\tau))\, d\tau \\ &- \mu_i^{\frac{1}{\alpha}-1} E_\alpha(\mu_i t^\alpha) \int_0^\infty \exp\left(-\mu_i^{\frac{1}{\alpha}} \tau\right) h^i(\xi(\tau))\, d\tau \end{aligned} \tag{5.33}$$

for $i = 1, \ldots, k^*$ and

$$(\mathcal{T}\xi)^i(t) := \int_0^t (t - \tau)^{\alpha-1} E_{\alpha,\alpha}(\mu_i(t - \tau)^\alpha) h^i(\xi(\tau))\, d\tau \tag{5.34}$$

for $i = k^* + 1, \ldots, d$, where $h(x) = (h^1(x), \ldots, h^d(x))^{\mathrm{T}}$ is the coordinate representation of the vector $h(x)$.

For $i = 1, \ldots, k^*$, write μ_i in the form

$$\mu_i = r_i(\cos \phi_i + \iota \sin \phi_i),$$

where $\iota = \sqrt{-1}$, $r_i > 0$ is the modulus and $\phi_i \in (-\frac{\alpha\pi}{2}, \frac{\alpha\pi}{2})$ is the argument of the complex number $\mu_i \in \Lambda_\alpha^u$. Let

$$w := \frac{1}{3} \min_{1 \leq i \leq k} \left\{ r_i^{\frac{1}{\alpha}} \cos(\phi_i/\alpha) \right\} > 0, \tag{5.35}$$

and define a weighted norm on $C_\infty([0, \infty); \mathbb{C}^d)$ by

$$\|\xi\|_w := \sup_{t \geq 0} \|\xi(t)\| e^{-wt}.$$

Then $(C_\infty([0, \infty); \mathbb{C}^d), \|\cdot\|_w)$ is also a Banach space ($\|\cdot\|_w$ has the three properties of a norm on the space $C([0, \infty); \mathbb{C}^d)$, but it may take the value $+\infty$).

From (5.28) and (5.29) it follows that h is Lipschitz continuous on a neighbourhood of the origin. Hence, there exists $\hat{\varepsilon} > 0$ such that $L_h(R) < \infty$ for all $0 < R < \hat{\varepsilon}$, where $L_h(R)$ is defined in Assumption 5.6. Using the notation

$$B_{C_\infty}(0, \hat{\varepsilon}) := \left\{ \xi \in C([0, \infty); \mathbb{C}^d) : \|\xi\|_\infty \leq \hat{\varepsilon} \right\},$$

the following proposition gives an estimate for the operator $\mathcal{T}$.

Proposition 5.2 *Consider transformed system* (5.30). *Then, there exists a positive constant $K(\alpha, A)$ depending on $\sigma(A)$ such that*

$$\|\mathcal{T}\xi - \mathcal{T}\hat{\xi}\|_w \leq K(\alpha, A) \, L_h(\max\{\|\xi\|_\infty, \|\hat{\xi}\|_\infty\}) \|\xi - \hat{\xi}\|_w$$

for all $\xi, \hat{\xi} \in B_{C_\infty}(0, \hat{\varepsilon})$.

Proof Let $\xi, \hat{\xi} \in C_\infty([0, \infty); \mathbb{C}^d)$ be arbitrary. For any index $i \in \{1, \ldots, k^*\}$ and any $t \geq 0$, it follows from the definition (5.33) of the operator $\mathcal{T}$ that

$$|(\mathcal{T}\xi)^i(t) - (\mathcal{T}\hat{\xi})^i(t)|$$

$$\leq \int_0^t \left| (t-\tau)^{\alpha-1} E_{\alpha,\alpha}(\mu_i(t-\tau)^\alpha) - \mu_i^{\frac{1}{\alpha}-1} E_\alpha(\mu_i t^\alpha) e^{-\mu_i^{\frac{1}{\alpha}}\tau} \right| |h^i(\xi(\tau)) - h^i(\hat{\xi}(\tau))| \, d\tau$$

$$+ \int_t^\infty \left| \mu_i^{\frac{1}{\alpha}-1} E_\alpha(\mu_i t^\alpha) e^{-\mu_i^{\frac{1}{\alpha}}\tau} \right| |h^i(\xi(\tau)) - h^i(\hat{\xi}(\tau))| \, d\tau.$$

Hence,

$$
|(\mathcal{T}\xi)^i(t) - (\mathcal{T}\hat{\xi})^i(t)|e^{-wt}
$$

$$
\leq \int_0^t \left| (t-\tau)^{\alpha-1} E_{\alpha,\alpha}(\mu_i(t-\tau)^\alpha) - \mu_i^{\frac{1}{\alpha}-1} E_\alpha(\mu_i t^\alpha) e^{-\mu_i^{\frac{1}{\alpha}}\tau} \right|
$$

$$
\times L_h(\max\{\|\xi\|_\infty, \|\hat{\xi}\|_\infty\}) \|\xi(\tau) - \hat{\xi}(\tau)\| e^{-w\tau} \, d\tau
$$

$$
+ \int_t^\infty \left| \mu_i^{\frac{1}{\alpha}-1} E_\alpha(t^\alpha \mu_i) e^{-\mu_i^{\frac{1}{\alpha}}\tau} \right| e^{w(\tau-t)}
$$

$$
\times L_h(\max\{\|\xi\|_\infty, \|\hat{\xi}\|_\infty\}) \|\xi(\tau) - \hat{\xi}(\tau)\| e^{-w\tau} \, d\tau.
$$

Since $\mu_i \in \Lambda_\alpha^u$, according to Lemma 5.7, we can find a constant $K(\alpha, \mu_i) > 0$ such that for all $t \geq 0$

$$
\int_0^t \left| (t-\tau)^{\alpha-1} E_{\alpha,\alpha}(\mu_i(t-\tau)^\alpha) - \mu_i^{\frac{1}{\alpha}-1} E_\alpha(\mu_i t^\alpha) e^{-\mu_i^{\frac{1}{\alpha}}\tau} \right| d\tau \leq K(\alpha, \mu_i). \quad (5.36)
$$

Furthermore, by definition of w as in (5.35) and Lemma 5.6 there exist $t_0 > 0$ and $M(\alpha, \mu_i) > 0$ such that for any $t \geq t_0$

$$
\int_t^\infty \left| \mu_i^{\frac{1}{\alpha}-1} E_\alpha(\mu_i t^\alpha) e^{-\mu_i^{\frac{1}{\alpha}}\tau} \right| e^{w(\tau-t)} \, d\tau
$$

$$
\leq r_i^{\frac{1}{\alpha}-1} \int_t^\infty \left(\frac{1}{\alpha} \exp\left(r_i^{\frac{1}{\alpha}} t \cos(\phi_i/\alpha) \right) + M(\alpha, \mu_i) t_0^{-\alpha} \right) \left| e^{-\mu_i^{\frac{1}{\alpha}}\tau} \right| e^{w(\tau-t)} \, d\tau
$$

$$
\leq r_i^{\frac{1}{\alpha}-1} \left(\frac{1}{\alpha} + M(\alpha, \mu_i) t_0^{-\alpha} \right) \int_0^\infty e^{-w\tau} \, d\tau, \qquad (5.37)
$$

while for $0 \leq t \leq t_0$,

$$
\int_t^\infty \left| \mu_i^{\frac{1}{\alpha}-1} E_\alpha(\mu_i t^\alpha) e^{-\mu_i^{\frac{1}{\alpha}}\tau} \right| e^{w(\tau-t)} \, d\tau \leq r_i^{\frac{1}{\alpha}-1} E_\alpha(r_i t_0^\alpha) \int_0^\infty e^{-w\tau} \, d\tau. \quad (5.38)
$$

From (5.36), (5.37), and (5.38), for each $i = 1, \ldots, k^*$, there exists a constant $\hat{K}(\alpha, \mu_i) > 0$ such that

$$
\|(\mathcal{T}\xi)^i - (\mathcal{T}\hat{\xi})^i\|_w \leq \hat{K}(\alpha, \mu_i) \, L_h(\max\{\|\xi\|_\infty, \|\hat{\xi}\|_\infty\}) \|\xi - \hat{\xi}\|_w.
$$

On the other hand, for $i = k^* + 1, \ldots, d$, using (5.34), for any $t \geq 0$,

$$
|(\mathcal{T}\xi)^i(t) - (\mathcal{T}\hat{\xi})^i(t)|e^{-wt} \leq \int_0^t (t-\tau)^{\alpha-1} |E_{\alpha,\alpha}(\mu_i(t-\tau)^\alpha)| e^{w(\tau-t)}
$$

$$
\times L_h(\max\{\|\xi\|_\infty, \|\hat{\xi}\|_\infty\}) \|\xi(\tau) - \hat{\xi}(\tau)\| e^{-w\tau} \, d\tau,
$$

which implies that

$$|(\mathcal{T}\xi)^i(t) - (\mathcal{T}\hat{\xi})^i(t)|e^{-wt} \leq \int_0^t \tau^{\alpha-1}|E_{\alpha,\alpha}(\mu_i\tau^\alpha)|e^{-w\tau}\,d\tau$$
$$\times L_h(\max\{\|\xi\|_\infty, \|\hat{\xi}\|_\infty\})\|\xi(\tau) - \hat{\xi}(\tau)\|_w.$$

Note for $\mu_i \in \sigma(A) \setminus \Lambda_\alpha^u$, it is clear that $|E_{\alpha,\alpha}(\mu_i t^\alpha)|$ is bounded on $[0, \infty)$. Hence

$$\int_0^\infty \tau^{\alpha-1}|E_{\alpha,\alpha}(\mu_i\tau^\alpha)|e^{-w\tau}\,d\tau < \infty.$$

Hence, for $i = k^* + 1, \ldots, d$,

$$\|(\mathcal{T}\xi)^i - (\mathcal{T}\hat{\xi})^i\|_w \leq \int_0^\infty \tau^{\alpha-1}|E_{\alpha,\alpha}(\mu_i\tau^\alpha)|e^{-w\tau}d\tau \times L_h(\max\{\|\xi\|_\infty, \|\hat{\xi}\|_\infty\})\|\xi - \hat{\xi}\|_w.$$

Let $\hat{K}(\alpha, \Lambda_\alpha^u) := \max_{1\leq i\leq k^*}\hat{K}(\alpha, \mu_i)$ and

$$K(\alpha, A) := \max\left\{\hat{K}(\alpha, \Lambda_\alpha^u), \max_{k^*+1\leq i\leq d}\int_0^\infty \tau^{\alpha-1}|E_{\alpha,\alpha}(\mu_i\tau^\alpha)|e^{-w\tau}\,d\tau\right\}.$$

This completes the proof of Proposition 5.2. $\square$

Henceforth, choose and fix the parameter $\gamma > 0$ in the definition of the transformation P above such that $\gamma < \frac{1}{2K(\alpha,A)}$. By this choice of γ, the system (5.27) and (5.30) are completely specified. Due to (5.29), there exists a positive constant ε such that

$$0 < \varepsilon < \hat{\varepsilon}, \quad \text{and} \quad K(\alpha, A)\,L_h(\varepsilon) \leq \frac{2}{3}. \tag{5.39}$$

Hence (5.39) and Proposition 5.2 yield:

Proposition 5.3 *For all $\xi, \hat{\xi} \in B_{C_\infty}(0, \varepsilon)$, where ε is a positive number satisfying* (5.39),

$$\|\mathcal{T}\xi - \mathcal{T}\hat{\xi}\|_w \leq \frac{2}{3}\,\|\xi - \hat{\xi}\|_w.$$

Step 3: Finishing of the proof of Theorem 5.3

It suffices to prove the instability for the zero solution of transformed system (5.30).

Assume to the contrary that the zero solution of (5.30) is stable. Let $\varepsilon > 0$ satisfy (5.39) and assume that the function g is Lipschitz continuous on $B_{C^d}(0, \varepsilon)$. Choose $\delta = \delta(\varepsilon) > 0$ such that for any $x_0 \in B_{C^d}(0, \delta)$ the solution $\phi(\cdot, x_0)$ satisfies $\|\phi(t, x_0)\| \leq \varepsilon$ for every $t \geq 0$.

Take and fix a vector $x_0 = (x_0^1, \ldots, x_0^{k^*}, 0, \ldots, 0)^{\mathrm{T}} \in B_{\mathbb{C}^d}(0, \delta) \setminus \{0\}$. By the variation of constants formula for Caputo FDES (see [6, Theorem 1] and [1, Remark 7.1, pp. 135–136]), $\phi(\cdot, x_0)$ has the representation

$$\phi(t, x_0) = E_\alpha(t^\alpha \mathrm{diag}(\mu_1, \ldots, \mu_d))x_0 + \int_0^t (t - \tau)^{\alpha-1} E_{\alpha,\alpha}((t - \tau)^\alpha \Lambda)h(\phi(\tau, x_0)) \, d\tau$$

$$=: (\phi^1(t, x_0), \phi^2(t, x_0), \ldots, \phi^d(t, x_0))^{\mathrm{T}}.$$

Writing this expression component-wise gives

$$\phi^i(t, x_0) = E_\alpha(\mu_i t^\alpha)x_0^i + \int_0^t (t - \tau)^{\alpha-1} E_{\alpha,\alpha}(\mu_i(t - \tau)^\alpha)h^i(\phi(\tau, x_0)) \, d\tau$$

$$= E_\alpha(\mu_i t^\alpha)\left[x_0^i + \int_0^t \frac{(t - \tau)^{\alpha-1} E_{\alpha,\alpha}(\mu_i(t - \tau)^\alpha)h^i(\phi(\tau, x_0))}{E_\alpha(\mu_i t^\alpha)} \, d\tau \right]$$

for $i = 1, \ldots, k^*$ and

$$\phi^i(t, x_0) = \int_0^t (t - \tau)^{\alpha-1} E_{\alpha,\alpha}(\mu_i(t - \tau)^\alpha)h^i(\phi(\tau, x_0)) \, d\tau$$

for $i = k^* + 1, \ldots, d$. By Lemma 5.6, $\lim_{t \to \infty} |E_\alpha(\mu_i t^\alpha)| = \infty$ for $i = 1, \ldots, k^*$. Therefore, $\sup_{t \in \mathbb{R}_{\geq 0}} \|\phi(t, x_0)\| \leq \varepsilon$ implies that

$$x_0^i = -\lim_{t \to \infty} \int_0^t \frac{(t - \tau)^{\alpha-1} E_{\alpha,\alpha}(\mu_i(t - \tau)^\alpha)h^i(\phi(\tau, x_0))}{E_\alpha(\mu_i t^\alpha)} \, d\tau$$

$$= -\mu_i^{\frac{1}{\alpha}-1} \int_0^\infty e^{-\mu_i^{\frac{1}{\alpha}}\tau} h^i(\phi(\tau, x_0)) \, d\tau$$

for $i = 1, \ldots, k^*$, using Lemma 5.8 for the limit.

Hence $\phi(\cdot, x_0)$ is a fixed point of the operator $\mathcal{T}$, i.e., $\mathcal{T}\phi(\cdot, x_0) = \phi(\cdot, x_0)$. Obviously $\mathcal{T}0 = 0$, and both the function $\phi(\cdot, x_0)$ and the zero function 0 belong to $B_{C_\infty}(0, \varepsilon)$. Thus, by Proposition 5.3,

$$\|\phi(\cdot, x_0) - 0\|_w = \|\mathcal{T}\phi(\cdot, x_0) - \mathcal{T}0\|_w$$

$$\leq K(\alpha, A)\, L_h(\varepsilon)\|\phi(\cdot, x_0) - 0\|_w \leq \frac{2}{3}\, \|\phi(\cdot, x_0) - 0\|_w,$$

which implies that $\phi(t, x_0) = 0$ for all $t \geq 0$.

This is a contradiction because $\phi(0, x_0) = x_0 \neq 0$. Thus, the zero solution of (5.30) is unstable. This completes the proof of Theorem 5.3. $\square$

5.5 Endnotes

Lemma 5.1 is from Cong, Tuan and Trinh [7, Lemma 5]. Mittag-Leffler asymptotically stable was called Mittag-Leffler stable in [7]. Due to the asymptotic behaviour of the Mittag-Leffler function this definition is equivalent to the definition of Mittag-Leffler stability used elsewhere in the literature, see [8–10].

The proof of Theorem 5.2 is taken from [7, Theorem 12], which extends [11, Theorem 3.1] from asymptotic to Mittag-Leffler asymptotic stability. This was possible due to the use of the weighted norm. The proof of Lemma 5.2 is from Cong, Tuan and Trinh [7, Example 35]. See also Theorem 6.2 for a proof of linearised stability using a Lyapunov function.

The proof of Theorem 5.3 is taken from [5, Theorem 5].

References

1. Diethelm, K.: The analysis of fractional differential equations, Springer Lecture Notes in Mathematics, vol. 2004. Springer, Heidelberg (2010)
2. Vergara, V., Zacher, R.: Optimal decay estimates for time-fractional and other nonlocal subdiffusion equations via energy methods. SIAM J. Math. Anal. **47**(1), 210–239 (2015)
3. Feng, Y., Li, L., Liu, J.G., Xu, X.: Continuous and discrete one dimensional autonomous fractional ODEs. Discr. Continuous Dyn. Syst. Ser. B **23**(8), 3109–3135 (2018)
4. Cong, N.D., Doan, T.S., Tuan, H.T.: Asymptotic stability of linear fractional systems with constant coefficients and small time dependent perturbations. Vietnam J. Math. **46**, 665–680 (2018)
5. Cong, N.D., Doan, T.S., Tuan, H.T.: An instability theorem of nonlinear fractional differential systems. Discr. Continuous Dyn. Syst. Ser. B **22**, 3079–3090 (2017)
6. Cong, N.D., Tuan, H.T.: Generation of nonlocal dynamical systems by fractional differential equations. J. Integral Equ. Appl. **29**, 585–608 (2017). https://doi.org/10.1216/JIE-2017-29-4-585
7. Cong, N.D., Tuan, H.T., Hieu, T.: On asymptotic properties of solutions to fractional differential equations. J. Math. Anal. Appl. **484**, 123759 (2020)
8. Li, Y., Chen, Y., Podlubny, I.: Mittag-Leffler stability of fractional order nonlinear dynamic systems. Automatica **45**, 1965–1969 (2009)
9. Li, Y., Chen, Y., Podlubny, I.: Stability of fractional-order nonlinear dynamic system: Lyapunov direct method and generalized Mittag-Leffler stability. Comput. Math. Appl. **59**, 1810–1821 (2010)
10. Stamova, I.: Mittag-Leffler stability of impulsive differential equations of fractional order. Q. Appl. Math. **73**, 239–244 (2015)
11. Cong, N.D., Doan, T.S., Siegmund, S., Tuan, H.T.: Linearized asymptotic stability for fractional differential equations. Electron. J. Qual. Theory Differ. Equ. **39**, 1–13 (2016)

Chapter 6
Lyapunov Functions

Abstract The f Mittag–Leffler stability of the trivial solution of nonlinear Caputo fractional differential equations is studied by the Lyapunov function method. The main tool is the Tuan–Trinh inequality from Chap. 2. The ultimate boundedness of solutions to dissipative fractional differential equations is also discussed.

Keywords Nonlinear Caputo fractional differential equations · Lyapunov function method · Mittag-Leffler stability · Ultimate boundedness

Lyapunov functions provide sufficient conditions for stability or asymptotic stability without explicit knowledge of the solutions of a Caputo FDEs. It is an alternative method to investigating such solutions to the linearised stability method used in Chap. 5. Moreover, it does not require knowledge about the spectral properties of the matrix in the linear part, if any, of such equations.

6.1 A Sufficient Condition for Stability

Let $\alpha \in (0, 1)$ and consider a nonlinear Caputo FDE

$$^{C}D_{0+}^{\alpha}x(t) = g(x(t)), \quad t > 0, \tag{6.1}$$

where $g : \mathbb{R}^d \to \mathbb{R}^d$ satisfies the following assumptions.

Assumption 6.1 $g(0) = 0$.

Assumption 6.2 For each $R > 0$, there exists $L_R > 0$ such that

$$\|g(x) - g(y)\| \le L_R \|x - y\| \quad \text{for all } x, y \in B_R(0). \tag{6.2}$$

"""

The global existence of solutions holds when g is locally Lipschitz and satisfies some kind of dissipativity condition as in the following theorem (see also Theorem 9.1).

Theorem 6.1 *Consider the Caputo FDE* (6.1). *Assume there is a function* $V : \mathbb{R}^d \to \mathbb{R}$ *satisfying*

(V1) *the function V is convex, differentiable on $\mathbb{R}^d$ and $V(0) = 0$;*

(V2) *there exist constants $a, b, C_1, C_2 > 0$ such that*

$$C_1 \|x\|^a \leq V(x) \leq C_2 \|x\|^b$$

  for all $x \in \mathbb{R}^d$;

(V3) *there are constants $C_3 \geq 0$ and $c \geq b$ such that*

$$\langle \nabla V(x), g(x) \rangle \leq -C_3 \|x\|^c$$

  for all $x \in \mathbb{R}^d$.

Then, the trivial solution of Caputo FDE (6.1) *is*

(a) *stable if $C_3 = 0$;*

(b) *Mittag-Lefffler asymptotically stable if $C_3 > 0$.*

Proof Consider a ball $B_R(0) = \{x \in \mathbb{R}^d : \|x\| \leq R\}$ and define

$$g_R(x) := \begin{cases} g(x), & \|x\| \leq R, \\ g(xR/\|x\|), & \|x\| \geq R. \end{cases}$$

This function satisfies a global Lipschitz condition. By Theorem 3.1, the Caputo FDE

$$^C D_{0+}^\alpha x(t) = g_R(x(t)), \quad t > 0 \tag{6.3}$$

has a unique global solution on $[0, \infty)$ for any initial condition. Choose $R_0 := \frac{1}{K}\left(\frac{C_1}{C_2}\right)^{1/b} R^{a/b}$, where $K > 1$ is large enough so that $R_0 < R$. Let $\|x_0\| < R_0$ and $\varphi(\cdot, x_0)$ be the solution of (6.3) on $[0, \infty)$ with the condition $\varphi(\cdot, x_0) = x_0$. Suppose by contradiction there is a time $t_0 > 0$ so that $\|\varphi(t_0, x_0)\| = R$. Denote $T^*(x_0) > 0$ as the first time for which $\varphi(T^*(x_0), x_0) = R$.

Due to the properties (V2) and (V3) are satisfied for the solution $\varphi(\cdot, x_0)$ in the time interval $[0, T^*(x_0)]$, it implies from (V3) and the Tuan-Trinh inequality (Theorem 2.1) that

$$^C D_{0+}^\alpha V(\varphi(t, x_0)) \leq \langle \nabla V(\varphi(t, x_0)), g(\varphi(t, x_0)) \rangle \leq 0, \qquad t \in [0, T^*(x_0)].$$

Hence, by Corollary 2.1,

$$V(\varphi(t, x_0)) \leq V(x_0), \quad \text{for all } t \in [0, T^*(x_0)]. \tag{6.4}$$

This combined with (V2) to give

$$\|\varphi(t, x_0)\| \leq \left(\frac{C_2}{C_1}\|x_0\|^b\right)^{1/a} \quad \text{for all } t \in [0, T^*(x_0)]. \tag{6.5}$$

Then, from (6.5),

$$\|\varphi(T^*(x_0), x_0)\| \leq \left(\frac{C_2}{C_1}\|x_0\|^b\right)^{1/a} \leq \left(\frac{C_2}{C_1}R_0^b\right)^{1/a} \leq \frac{1}{K^{b/a}}R < R,$$

a contradiction. Hence, the solution $\varphi(\cdot, x_0)$ with an initial condition $\|x_0\| < R_0$ remains in the ball $B_R(0)$ for all $t \geq 0$. It is thus also the unique global solution of the original Caputo FDE (6.1) on $[0, \infty)$.

(a) Consider the case $C_3 = 0$. For any $\varepsilon > 0$ small enough (for example, $\varepsilon < R$), choose $\delta := \frac{1}{K}\left(\frac{C_1}{C_2}\right)^{1/b}\varepsilon^{a/b}$, where $K > 1$ is large enough so that $\delta < \varepsilon$. Let $x_0 \in B_\delta$ be arbitrary. Then, from (6.5),

$$\|\varphi(t, x_0)\| \leq \left(\frac{C_2}{C_1}\|x_0\|^b\right)^{1/a} \leq \left(\frac{C_2}{C_1}\delta^b\right)^{1/a} < \varepsilon$$

for all $t \in [0, \infty)$. This shows that the trivial solution of the Caputo FDE (6.1) is stable.

(b) Assume that $C_3 > 0$. By using the same arguments as above, it follows that the trivial solution of the Caputo FDE (6.1) is stable. Hence, given $\varepsilon > 0$, there exists $\delta > 0$ such that every solution $\varphi(\cdot, x_0)$ of the Caputo FDE (6.1) with $\|x_0\| < \delta$ satisfies $\|\varphi(t, x_0)\| < \varepsilon$ for all $t \geq 0$.

By properties (V2)–(V3) and Theorem 2.1,

$$^C D_{0+}^\alpha V(\varphi(t, x_0)) \leq -C_3\|\varphi(t, x_0)\|^c \leq -\frac{C_3}{C_2^{c/b}}(V(\varphi(t, x_0)))^{c/b}, \quad \forall t \geq 0.$$

Write $A := -\frac{C_3}{C_2^{c/b}}$, $p := \frac{c}{b}$ and consider the following initial value problem

$$\begin{cases} ^C D_{0+}^\alpha y(t) = Ay^p(t), & t > 0, \\ y(0) = V(x_0) > 0. \end{cases} \tag{6.6}$$

Then $V(\varphi(\cdot, x_0))$ is a sub-solution of (6.6). Furthermore, define $C = V(x_0)t_1^{\frac{\alpha}{p}}$ and

$$t_1^\alpha = \frac{V(x_0)^{1-p}}{-A}\left(\frac{2^\alpha}{\Gamma(1-\alpha)} + \frac{\alpha}{p}\frac{2^{\alpha+\frac{\alpha}{p}}}{\Gamma(2-\alpha)}\right).$$

The function below

$$w(t) := \begin{cases} V(x_0), & t \in [0, t_1], \\ Ct^{-\frac{\alpha}{p}}, & t \geq t_1, \end{cases}$$

is a super-solution to (6.6) (see Vergara and Zacher [1, p. 333]).

Thus, by the comparison theorem (Theorem 2.6),

$$V(\varphi(t, x_0)) \leq w(t), \quad \forall t \geq 0.$$

This implies that for any $x_0 \in B_\delta(0) \setminus \{0\}$, that there exists a constant $d > 0$ such that

$$\|\varphi(t, x_0)\| \leq \left(\frac{1}{C_1}V(\varphi(t, x_0))\right)^{1/a} \leq \left(\frac{d}{C_1(1 + t^{\alpha/p})}\right)^{1/a}$$

for all $t \geq 0$.

Notice that from the existence and uniqueness of the solution to (6.1) in $B_\varepsilon(0)$, if $x_0 = 0$ then $\varphi(\cdot, 0) = 0$. Hence, the trivial solution to the original system (6.1) is Mittag-Lefffler stable. This completes the proof of Theorem 6.1. $\qquad\square$

One of the advantages of the Lyapunov function method is that it can be applied to systems where their fields do not have linear parts.

Example 6.1 Consider the scalar Caputo FDE

$$^C D_{0+}^\alpha x(t) = -x(t)^3 + h(x(t)), \quad t > 0, \tag{6.7}$$

where $h : \mathbb{R} \to \mathbb{R}$ is differentiable at the origin and satisfies

$$h(0) = 0, \quad \lim_{x \to 0} \frac{h(x)}{x^3} = 0.$$

Choose an $R > 0$ such that

$$2x(-x^3 + h(x)) \leq -x^4, \quad \forall x \in B_R(0).$$

Then the conditions of Theorem 6.1 are satisfied with $C_1 = C_2 = 1$, $a = b = 2$, $C_3 = 1$ and $c = 4$ for the Lyapunov function $V(x) = x^2$, $x \in \mathbb{R}$. Thus, the trivial solution of (6.7) is Mittag-Leffler asymptotically stable.

6.2 Linearised Stability for Nonlinear Caputo FDEs

Lyapunov functions allow a simpler proof for the linearised stability for nonlinear Caputo FDEs than that of Theorem 5.2 when all of the eigenvalues of the matrix in the linear part of the equation have negative real parts.

Consider an autonomous C^1 vector field $f : \mathbb{R}^d \to \mathbb{R}^d$ with a steady state x^* for the FDE, i.e., $f(x^*) = 0$, and suppose that the linear part $A := \nabla f(x^*)$ has all eigenvalues with negative real parts. This means zero solution of the that the corresponding linear ODE is asymptotically stable and by Lyapunov's Theorem there exists a real valued symmetric $d \times d$-matrix P such that

$$PA + A^T P = -I. \tag{6.8}$$

Translating the origin to x^* gives

$$g(z) := f(x^* + z) = Az + n(z),$$

where $n(0) = 0$, $\nabla n(0) = 0$, so $n(z) \sim 0(z^2)$. Hence, given $\epsilon > 0$ there is a $K(\epsilon) > 0$ such that

$$\langle z, Pn(z) \rangle + \langle Pn(z), z \rangle \leq K(\epsilon)\|z\|^2 \tag{6.9}$$

for $z^T P z \leq \epsilon$, i.e., lies in an ellipsoidal region about the origin, where $K(\epsilon) \to 0$ as $\epsilon \to 0$.

The proof of the next theorem uses the Lyapunov function $V(x) = x^T P x$ and the fact that there are positive constants $v^- < v^+$ such that

$$v^- x^T x \leq x^T P x \leq v^+ x^T x, \quad x \in \mathbb{R}^d. \tag{6.10}$$

Theorem 6.2 *Assume that all of the eigenvalues of the matrix A have negative real parts and that the nonlinear function n satisfies (6.9). Then the zero solution of the linear FDE*

$$^C D_{0+}^\alpha z(t) = Az(t) \tag{6.11}$$

is globally Mittag-Leffler asymptotically stable and the zero solution of the nonlinear Caputo FDE

$$^C D_{0+}^\alpha z(t) = g(z(t)) = Az(t) + n(z(t)) \tag{6.12}$$

is locally Mittag-Leffler asymptotically stable.

Proof From the Lyapunov Theorem $z^T (PA + A^T P)z \leq -z^T z$ for all $z \in \mathbb{R}^d$. Then by the Tuan-Trinh inequality (2.5), $V(z(t))$ along the solutions of the linear Caputo FDE (6.11) satisfies

$$^{C}D_{0+}^{\alpha}V(z(t)) \leq \langle \nabla V(z(t)), Az(t)) \rangle = z(t)^{T}\left(A^{T}P + PA\right)z(t)$$

$$= -z^{T}(t)z(t) \leq -\frac{1}{\nu^{+}}V(z(t)), \quad t > 0,$$

where (6.10) has been used in the final inequality. Hence, by Lemma 2.1,

$$V(z(t)) \leq V(z(0))E_{\alpha}\left(-\frac{1}{\nu^{+}}t^{\alpha}\right), \quad t \geq 0.$$

Using (6.10), this gives

$$\|z(t)\|^{2} \leq \frac{\nu^{+}}{\nu^{-}}\|z(0)\|^{2}E_{\alpha}\left(-\frac{1}{\nu^{+}}t^{\alpha}\right), \quad t \geq 0,$$

i.e., the zero solution is globally Mittag-Leffler asymptotically stable.

Similarly, by the Tuan-Trinh inequality (2.5), the solutions of the nonlinear Caputo FDE (6.12) satisfy

$$\begin{aligned}
^{C}D_{0+}^{\alpha}V(z(t)) &\leq \langle \nabla V(z(t)), g(z(t))\rangle \\
&\leq \langle \nabla V(z(t), Az(t)\rangle + \langle \nabla V(z(t), n(z(t))\rangle \\
&\leq -z^{T}(t)z(t) + \langle z(t), Pn(z(t))\rangle + \langle Pn(z(t)), z(t)\rangle.
\end{aligned}$$

Pick $\epsilon > 0$ small enough such that $1 - K(\epsilon) > \frac{1}{2}$. Then, by the inequality (6.9),

$$^{C}D_{0+}^{\alpha}V(z(t)) \leq -\|z(t)\|^{2} + K(\epsilon)\|z(t)\|^{2} \leq -\frac{1}{2}\|z(t)\|^{2} \leq -\frac{1}{2\nu^{+}}V(z(t)),$$

provided $V(z(t)) \leq \epsilon$ for $t \geq 0$ when the initial value $V(z(0)) \leq \epsilon$.

To show this, apply Lemma 2.3 to the Caputo differential inequality

$$^{C}D_{0+}^{\alpha}V(z(t)) \leq -\frac{1}{2\nu^{+}}V(z(t)) + \mu, \quad t > 0,$$

for an arbitrary $\mu > 0$ to obtain

$$V(z(t)) \leq V(z(0))E_{\alpha}\left(-\frac{1}{2\nu^{+}}t^{\alpha}\right) + 2\mu\nu^{+}\left(1 - E_{\alpha}\left(-\frac{1}{2\nu^{+}}t^{\alpha}\right)\right)$$

for $t \geq 0$. Thus, for $V(z(0)) \leq \epsilon$,

$$V(z(t)) \leq \epsilon E_{\alpha}\left(-\frac{1}{2\nu^{+}}t^{\alpha}\right) + 2\mu\nu^{+}\left(1 - E_{\alpha}\left(-\frac{1}{2\nu^{+}}t^{\alpha}\right)\right) \leq \epsilon + 2\mu\nu^{+}$$

for $t \geq 0$. Since $\mu > 0$ was arbitrary, letting it converge to zero gives $V(z(t)) \leq \epsilon$ for $t \geq 0$.

Thus, for solutions with initial value $V(z(0)) \leq \epsilon$, the following Caputo differential inequality

$$^{C}D_{0+}^{\alpha} V(z(t)) \leq -\frac{1}{2\nu^{+}} V(z(t)), \quad t > 0,$$

is valid. Hence

$$V(z(t)) \leq V(z(0)) E_{\alpha}\left(-\frac{1}{2\nu^{+}} t^{\alpha}\right) \to 0 \text{ as } t \to \infty,$$

from which it follows that

$$\|z(t)\|^{2} \leq \frac{\nu^{+}}{\nu^{-}} \|z(0)\|^{2} E_{\alpha}\left(-\frac{1}{2\nu^{+}} t^{\alpha}\right) \to 0 \text{ as } t \to \infty.$$

This gives the local Mittag-Leffler asymptotical stability of the nonlinear Caputo FDE (6.12). $\qquad\square$

6.3 Ultimate Boundedness

The solutions $\varphi(\cdot, x_0)$ of the Caputo FDE (6.1) on $\mathbb{R}^d$ are said to be *ultimately bounded* if there exists an $R_0 > 0$ and for each bounded subset D of $\mathbb{R}^d$ there exists $T_D > 0$ such that

$$\|\varphi(t, x_0)\| \leq R_0 \quad \text{for all } t \geq T_D, x_0 \in D.$$

Essentially, the ball $B_{R_0}(0)$ in $\mathbb{R}^d$ of radius R_0 about the origin is an absorbing set of the Caputo FDE (6.1).

Theorem 9.2 gives the existence of such an absorbing set when the vector field g satisfies the dissipativity condition (9.3) using the Lyapunov function $V(x) = \|x\|^2$. The next result uses a more general Lyapunov function.

Theorem 6.3 *Consider the Caputo FDE (6.1). Assume there is a function $V : \mathbb{R}^d \to \mathbb{R}$ satisfying*

(V1) the function V is convex, differentiable on $\mathbb{R}^d$ and $V(0) = 0$;

(V2) there exist constants $a, b, C_1, C_2 > 0$ such that

$$C_1 \|x\|^a \leq V(x) \leq C_2 \|x\|^b$$

for all $x \in \mathbb{R}^d$;

(V3)'   there are constants $C_3 > 0$ and $C_4 \geq 0$ such that

$$\langle \nabla V(x), g(x) \rangle \leq C_4 - C_3 \|x\|^b$$

for all $x \in \mathbb{R}^d$.

Then, the solutions of the Caputo FDE (6.1) is ultimately bounded.

Proof The global existence and uniqueness of solutions of the Caputo FDE (6.1) follows by a similar argument to that in the proofs of Theorems 9.1 and 6.1.

Let $\varphi(\cdot, x_0)$ be the solution of the Caputo FDE (6.1) for $x_0 \in D$, an arbitrary bounded subset of $\mathbb{R}^d$. By the property (V3)' and Theorem 2.1,

$$^C D_{0+}^\alpha V(\varphi(t, x_0)) \leq C_4 - C_3 \|\varphi(t, x_0))\|^b, \ t > 0.$$

Due to (V2),

$$- C_3 \|\varphi(t, x_0))\|^b \leq -\frac{C_3}{C_2} V(\varphi(t, x_0)),$$

so

$$^C D_{0+}^\alpha V(\varphi(t, x_0)) \leq C_4 - \frac{C_3}{C_2} V(\varphi(t, x_0)), \ t > 0.$$

Hence, by Lemma 2.3, for all $t \geq 0$,

$$V(\varphi(t, x_0)) \leq V(x_0) E_\alpha(-\gamma t^\alpha) + \frac{C_2 C_4}{C_3} \left(1 - E_\alpha(-\gamma t^\alpha)\right),$$

where $\gamma := \frac{C_3}{C_2}$. This implies

$$V(\varphi(t, x_0)) \leq C_2 R_D^b E_\alpha(-\gamma t^\alpha) + \frac{C_2 C_4}{C_3} \left(1 - E_\alpha(-\gamma t^\alpha)\right),$$

since, by Property (V2), $V(x_0) \leq C_2 \|x_0\|^b \leq C_2 R_D^b$ for $x_0 \in D$. Thus,

$$V(\varphi(t, x_0)) \leq 1 + \frac{C_2 C_4}{C_3} =: R_*, \ t \geq T_D,$$

where T_D is the time t with

$$C_2 R_D^b E_\alpha(-\gamma t^\alpha) = 1$$

(T_D exists and is unique since $E_\alpha(-\gamma t^\alpha)$ is positive and strictly decreasing). Finally, by (V2) again,

$$C_1 \|\varphi(t, x_0)\|^a \leq V(\varphi(t, x_0)),$$

so

$$\|\varphi(t, x_0)\| \leq \left(\frac{1}{C_1}\left(1 + \frac{C_2 C_4}{C_3}\right)\right)^{1/a} \leq \left(\frac{R_*}{C_1}\right)^{1/a} =: R_0, \ t \geq T_D.$$

6.4 Endnotes

An earlier version of Theorem 6.1 was given by Tuan and Trinh [2, Theorem 3]. It was improved and strengthened by Cong, Tuan and Trinh [3, Theorem 28]. Theorem 6.2 is adapted from Kloeden [4, Theorem 2]. Theorem 6.3 is new.

References

1. Vergara, V., Zacher, R.: Optimal decay estimates for time-fractional and other nonlocal subdiffusion equations via energy methods. SIAM J. Math. Anal. **47**(1), 210–239 (2015)
2. Tuan, H.T., Trinh, H.: Stability of fractional-order nonlinear systems by Lyapunov direct method. IET Control Theory Appl. **12**, 2417–2422 (2018)
3. Cong, N.D., Tuan, H.T., Hieu, T.: On asymptotic properties of solutions to fractional differential equations. J. Math. Anal. Appl. **484**, 123759 (2020)
4. Kloeden, P.E.: An elementary inequality for dissipative Caputo fractional differential equations. Fract. Calc. Appl. Anal. **26**, 2166–2174 (2023) https://doi.org/10.1007/s13540-023-00192-x

Chapter 7
Non-intersecting Solutions of Scalar Caputo FDEs

Abstract The separability of solutions to Caputo fractional differential equations is investigated. In particular, separability is confirmed in the scalar case and for triangular systems. Furthermore, both upper and lower estimates for separability are established. A counterexample shows separability does not usually hold in the general multidimensional setting.

Keywords Caputo fractional differential equations · Separability of solutions · Upper estimate · Lower estimate · Intersection of trajactories

There is an extensive discussion on the non-intersecting solutions of scalar Caputo FDEs in the literature, see [1] and the references cited therein. As for scalar ODEs any two trajectories of solutions either coincide or do not intersect each other. This also holds for higher dimensional triangular systems of FDEs, but not, in general, for other higher dimensional systems of FDEs where two different trajectories can meet.

Consider the scalar Caputo FDE

$$^{C}D_{0+}^{\alpha}x(t) = g(x(t)), \quad t > 0, \tag{7.1}$$

where $g : \mathbb{R} \to \mathbb{R}$ satisfies the global Lipschitz condition

$$|g(x) - g(y)| \le L|x - y|, \quad x, y \in \mathbb{R}. \tag{7.2}$$

Theorem 7.1 *Suppose that g satisfies the Lipschitz condition* (7.2). *Then, for any two different initial values $x_{10} \ne x_{20}$ in $\mathbb{R}$, the trajectories of the corresponding solutions of the Caputo FDE* (7.1) *do not meet, i.e.,*

$$\varphi(t, x_{10}) \ne \varphi(t, x_{20}) \quad \textit{for all } t \ge 0.$$

Proof For definiteness assume that

$$\varphi(0, x_{10}) = x_{10} < x_{20} = \varphi(0, x_{20}) \tag{7.3}$$

T. S. Doan et al., *Attractors of Caputo Fractional Differential Equations*,
SpringerBriefs in Mathematics, https://doi.org/10.1007/978-3-032-05511-8_7

and fix a time interval $[0, T]$ with a $T > 0$ is arbitrary. To prove the theorem it needs to be shown that $\varphi(t, x_{10}) < \varphi(t, x_{20})$ for all $t \in [0, T]$. Suppose that this is not true, then there is a $t_1 \in (0, T]$ such that $\varphi(t_1, x_{10}) = \varphi(t_1, x_{20})$. By the continuity property of the solutions and (7.3), there is $t^* \in (0, T]$ such that

$$\varphi(t^*, x_{10}) = \varphi(t^*, x_{20}) \tag{7.4}$$

and

$$\varphi(t, x_{10}) < \varphi(t, x_{20}) \quad \text{for all } t \in [0, t^*). \tag{7.5}$$

Define

$$n(x) := g(x) + Lx \quad \text{for all } x \in \mathbb{R}, \tag{7.6}$$

where $L \geq 0$ is a constant satisfied the condition (7.2). Notice that $n : \mathbb{R} \to \mathbb{R}$ is nondecreasing in x, since, by (7.2) and (7.6), for any $x \leq y$

$$n(y) - n(x) = L(y - x) + g(y) - g(x) \geq (L - L)(y - x) = 0.$$

Then the FDE (7.1) is written in the form

$$^C D_{0+}^\alpha x(t) = -Lx(t) + n(x(t)), \quad t > 0.$$

By the variation of constants formula (Lemma 3.5), the solutions $\varphi(\cdot, x_{10})$ and $\varphi(\cdot, x_{20})$ of (7.1) on the interval $[0, t^*]$ satisfy

$$\varphi(t, x_{10}) = E_\alpha(-Lt^\alpha)x_{10}$$
$$+ \int_0^t (t - \tau)^{\alpha-1} E_{\alpha,\alpha}(-L(t - \tau)^\alpha)n(\varphi(\tau, x_{10}))d\tau, \tag{7.7}$$
$$\varphi(t, x_{20}) = E_\alpha(-Lt^\alpha)x_{20}$$
$$+ \int_0^t (t - \tau)^{\alpha-1} E_{\alpha,\alpha}(-L(t - \tau)^\alpha)n(\varphi(\tau, x_{20}))d\tau. \tag{7.8}$$

Due to $E_{\alpha,\alpha}(s) > 0$ for all $s \in \mathbb{R}$ (see, e.g., [2, Corolary 3.7, p. 29]) and $E_{\alpha,\alpha}(-L(t - \tau)^\alpha)n(x)$ is nondecreasing in the variable x for fixed each $0 \leq \tau \leq t \leq t^*$. Hence, by combining (7.4) and (7.7)–(7.8), it follows $\varphi(t^*, x_{10}) < \varphi(t^*, x_{20})$, a contradiction with (7.4). Consequently, the conclusion of the theorem is true. $\qquad\square$

The previous result also holds when the vector field g satisfies a local Lipschitz condition. Let $-\infty \leq a < b \leq \infty$ and $g : (a, b) \to \mathbb{R}$ be locally Lipschitz continuous, i.e., for any compact interval $K \subset (a, b)$ there exists a positive constant L_K such that

$$|g(x) - g(y)| \leq L_K|x - y|, \quad \forall x, y \in K. \tag{7.9}$$

Theorem 7.2 [3, Theorem 36] *Assume that the function g satisfies the condition (7.9). Then for any pair of distinct points $x_1, x_2 \in (a, b)$, the solutions of the scalar Caputo FDE (7.1) starting from x_1 and x_2, respectively, do not meet.*

Proof By the local and uniqueness existence and uniqueness theorem, [3, Theorem 4], for $x_i \in (a, b), i = 1, 2$, the initial value problem (7.1), $x(0) = x_i$, has the unique solution denoted by $\varphi(\cdot, x_i)$ on the maximal interval of existence $[0, T_b(x_i))$.

Without loss of generality let $x_1 < x_2$ and assume that $\varphi(\cdot, x_1)$ and $\varphi(\cdot, x_2)$ meet at some $t \in (0, T_b(x_1)) \cap (0, T_b(x_2))$. Let $t_1 := \inf\{t \in (0, T_b(x_1)) \cap (0, T_b(x_2)) : \varphi(t, x_1) = \varphi(t, x_2)\}$. It is obvious that $0 < t_1 < \min\{T_b(x_1), T_b(x_1)\}$ and

$$\varphi(t_1, x_1) = \varphi(t_1, x_2), \quad \varphi(t, x_1) < \varphi(t, x_2), \quad \forall t \in [0, t_1).$$

Take $r_1, r_2 > 0$ such that $[x_1 - r_1, x_2 + r_2] \subset (a, b)$ and $\varphi(t, x_1), \varphi(t, x_2) \in [x_1 - r_1, x_2 + r_2]$ for all $t \in [0, t_1]$. Then, by the local Lipschitz assumption (7.9) of g, the restricted function

$$g_1(x) := \begin{cases} g(x), & \text{if } x \in [x_1 - r_1, x_2 + r_2], \\ g(x_2 + r_2), & \text{if } x > x_2 + r_2, \\ g(x_1 - r_1), & \text{if } x < x_1 - r_1, \end{cases}$$

is globally Lipschitz continuous on $\mathbb{R}$, so by Theorem 3.2 the Caputo FDE

$$^C D_{0+}^\alpha x(t) = g_1(x(t)), \quad x(0) = x_i, \ i = 1, 2, \tag{7.10}$$

has unique solutions $\tilde{\varphi}(\cdot, x_i)$ on $[0, \infty)$ for each $i = 1, 2$. On the other hand, using Theorem 7.1,

$$\tilde{\varphi}(t, x_1) < \tilde{\varphi}(t, x_2), \quad \forall t \geq 0.$$

However, $\varphi(t, x_1), \varphi(t, x_2) \in [x_1 - r_1, x_2 + r_2]$ holds for all $t \in [0, t_1]$, so

$$^C D_{0+}^\alpha \varphi(t, x_i) = g(\varphi(t, x_i)) = g_1(\varphi(t, x_i)), \quad t \in (0, t_1], \ i = 1, 2.$$

This implies that

$$\varphi(t, x_1) = \tilde{\varphi}(t, x_1) < \tilde{\varphi}(t, x_2) = \varphi(t, x_2)$$

for all $t \in [0, t_1]$, which is a contradiction. Thus two solutions $\varphi(\cdot, x_1)$ and $\varphi(\cdot, x_2)$ do not meet and the proof is complete. $\square$

7.1 Bounds for Solutions of Scalar Caputo FDEs

Upper and lower bounds for solutions of scalar Caputo FDEs allow a better understanding of geometry of solutions of such FDEs. The lower separation bound is particularly important as it shows that two solutions cannot intersect.

Theorem 7.3 *Assume that g satisfies the Lipschitz condition (7.2). For any two solutions $x_1(\cdot)$, $x_2(\cdot)$ of the FDE (7.1) and any $t \in [0, T]$, the following estimate holds*

$$|x_2(t) - x_1(t)| \geq |x_2(0) - x_1(0)|\, E_\alpha\big(-Lt^\alpha\big).$$

Proof For definiteness, assume $x_2(0) \geq x_1(0)$. Then, by Theorem 7.1, $x_2(t) \geq x_1(t)$ for any $t \in [0, T]$. For a fixed but arbitrary $t \in [0, T]$, repeating the arguments of the proof of Theorem 7.1 on the interval $[0, t]$ yields

$$x_2(t) - x_1(t) = E_\alpha(-Lt^\alpha)(x_2(0) - x_1(0))+$$

$$+ \int_0^t (t - \tau)^{\alpha-1} E_{\alpha,\alpha}(-L(t-\tau)^\alpha)(n(x_2(\tau)) - n(x_1(\tau)))d\tau$$

$$\geq E_\alpha(-Lt^\alpha)(x_2(0) - x_1(0)).$$

This completes the proof. $\square$

Corollary 7.1 *Assume that g satisfies the Lipschitz condition (7.2) and that $g(0) = 0$ for all $x \in \mathbb{R}$. Then, for any solution $x(\cdot)$ of the scalar Caputo FDE (7.1),*

$$|x(t)| \geq |x(0)|\, E_\alpha\big(-Lt^\alpha\big)$$

for $t \in [0, T]$.

Proof Since $g(0) = 0$, the FDE (7.1) has trivial solution. The conclusion of the corollary is obtained by applying Theorem 7.3 for the solution $x(\cdot)$ and the trivial solution of (7.1). $\square$

For the divergence rate and upper bound for solutions of the FDEs the following statements are easy modifications of well known results.

Theorem 7.4 *Assume that g satisfies the Lipschitz condition (7.2). Then, for any two solutions $x_1(\cdot)$, $x_2(\cdot)$ of the scalar Caputo FDE (7.1),*

$$|x_2(t) - x_1(t)| \leq |x_2(0) - x_1(0)|\, E_\alpha\big(Lt^\alpha\big)$$

for $t \in [0, T]$.

Proof By Lemma 1.1, for all $t \in [0, T]$,

$$x_1(t) = x_1(0) + \frac{1}{\Gamma(\alpha)} \int_0^t (t - \tau)^{\alpha-1} g(x_1(\tau))d\tau,$$

$$x_2(t) = x_2(0) + \frac{1}{\Gamma(\alpha)} \int_0^t (t - \tau)^{\alpha-1} g(x_2(\tau))d\tau.$$

Hence,

$$|x_2(t) - x_1(t)| \le |x_2(0) - x_1(0)| +$$

$$+ \frac{1}{\Gamma(\alpha)} \int_0^t (t - \tau)^{\alpha-1} (g(x_2(\tau)) - g(x_1(\tau))) d\tau$$

$$\le |x_2(0) - x_1(0)| + \frac{1}{\Gamma(\alpha)} \int_0^t (t - \tau)^{\alpha-1} L |x_2(\tau) - x_1(\tau)| d\tau$$

$$= |x_2(0) - x_1(0)| + \frac{L}{\Gamma(\alpha)} \int_0^t (t - \tau)^{\alpha-1} |x_2(\tau) - x_1(\tau)| d\tau, \quad t \in [0, T].$$

Then, by the Gronwall inequality (Lemma 2.4),

$$|x_2(t) - x_1(t)| \le |x_2(0) - x_1(0)| \, E_\alpha(L t^\alpha), \quad t \in [0, T]. \qquad \square$$

Corollary 7.2 *Assume that g satisfies the Lipschitz condition (7.2) and that $g(0) = 0$. Then, for any solution $x(\cdot)$ of the scalar Caputo FDE (7.1)*

$$|x(t)| \le |x(0)| \, E_\alpha(L t^\alpha)$$

for $t \in [0, T]$.

Proof Since $g(0) = 0$, the FDE (7.1) has trivial solution. The conclusion of the corollary is obtained by applying Theorem 7.4 for the solution $x(\cdot)$ and the trivial solution of (7.1). $\qquad \square$

Remark 7.1 It is easily seen that Theorem 7.4 and Corollary 7.2 also hold for high dimensional systems of Caputo FDEs.

7.2 Triangular Systems of FDEs

Consider a d-dimensional triangular system of Caputo FDEs

$$\begin{cases} {}^C D_{0+}^\alpha x_1(t) & = g_1(x_1(t)), \\ {}^C D_{0+}^\alpha x_2(t) & = g_2(x_1(t), x_2(t)), \\ \quad \cdots \\ {}^C D_{0+}^\alpha x_d(t) & = g_d(x_1(t), x_2(t), \cdots, x_d(t)), \end{cases} \tag{7.11}$$

where $x(\cdot) = \big(x_1(\cdot), \ldots, x_d(\cdot)\big)^{\mathrm{T}} \in \mathbb{R}^d$ and $g = (g_1, \ldots, g_d)^{\mathrm{T}}$ satisfies the global Lipschitz condition with a Lipschitz constant L, i.e., component-wise

$$|g_i(x_1, \ldots, x_i) - g_i(y_1, \ldots, y_i)| \le L\sqrt{(x_1 - y_1)^2 + \cdots (x_i - y_i)^2}, \quad i = 1, \ldots, d. \tag{7.12}$$

This triangular system can be solved successively coordinate-wise and reduces to solving successively one-dimensional FDEs. Hence the triangular system inherits many features of one-dimensional FDEs.

Proposition 7.1 *Assume that the Lipschitz condition (7.12) is satisfied. Then, for any two solutions $x(\cdot)$, $y(\cdot)$ of the triangular Caputo FDE (7.11)*

$$\|x(t) - y(t)\| \geq \|x(0) - y(0)\| E_\alpha\big(-Lt^\alpha\big)$$

for $t \in [0, T]$.

Proof Let $x(\cdot) = (x_1(\cdot), \ldots, x_d(\cdot))^{\mathrm{T}}$ and $y(\cdot) = (y_1(\cdot), \ldots, y_d(\cdot))^{\mathrm{T}}$ be arbitrary two solutions of the triangular Caputo FDE (7.11). Consider the first equation in (7.11). It is a one-dimensional equation for the first coordinate. Applying Theorem 7.3 to this equation gives

$$|x_1(t) - y_1(t)| \geq |x_1(0) - y_1(0)| E_\alpha\big(-Lt^\alpha\big), \quad t \in [0, T].$$

Since the first coordinate is solvable from the first equation its solution can be substituted into the second equation of the system (7.11) to yield a one-dimensional FDE for the second coordinate

$$^C D_{0+}^\alpha u(t) = g_2(x_1(t), u(t)) =: \hat{g}_2(t, u(t)),$$

where due to (7.12) the function $\hat{g}_2(\cdot, \cdot) : [0, T] \times \mathbb{R} \to \mathbb{R}$ is Lipschitz continuous in the second variable uniformly with the Lipschitz constant L. Applying Theorem 7.3 (the proof of which remains valid here) to the solutions $x_2(\cdot)$ and $y_2(\cdot)$ of this scalar Caputo FDE gives

$$|x_2(t) - y_2(t)| \geq |x_2(0) - y_2(0)| E_\alpha\big(-Lt^\alpha\big), \quad t \in [0, T].$$

Continue this procedure gives

$$|x_i(t) - y_i(t)| \geq |x_i(0) - y_i(0)| E_\alpha\big(-Lt^\alpha\big) \tag{7.13}$$

for any $i = 1, \ldots, d$ and each $t \in [0, T]$. $\square$

An important particular case of the triangular system of FDEs (7.11) is a linear triangular system of Caputo FDEs

$$^C D_{0+}^\alpha x(t) = Ax(t), \tag{7.14}$$

where $x \in \mathbb{R}^d$ and A is a triangular $(d \times d)$-matrix, i.e., $A = \big(a_{i,j}\big)_{1 \leq i,j \leq d}$ with either $a_{i,j} = 0$ for all $i > j$ (upper triangular) or $a_{i,j} = 0$ for all $i < j$ (lower triangular). Moreover, there exists a constant L such that

$$\|A\| \leq L.$$

Clearly, Proposition 7.1 is applicable to the linear triangular system (7.14).

Proposition 7.2 *Assume that the triangular matrix A. Then, for any solution $x(\cdot)$ of the triangular linear Caputo FDE (7.14)*

$$\|x(t)\| \geq \|x(0)\| \, E_\alpha(-Lt^\alpha)$$

for all $t \in [0, T]$.

Proof The system (7.14) has trivial solution 0, so Proposition 7.1 can be applied to the two solution $x(\cdot)$ and 0 of (7.14).   □

7.3 Counterexample in Higher Dimensional FDEs

Different trajectories of a Caputo FDE may intersect each other in the high dimensional case.

Theorem 7.5 *For any $d \geq 2$ there exists a system of Caputo FDEs with a property that there two different solutions $x_1(\cdot)$, $x_2(\cdot)$ with $x_1(0) \neq x_2(0)$ which intersect each other at some finite time $0 < T < \infty$, i.e., $x_1(T) = x_2(T)$.*

Proof It suffices to construct a two-dimensional system with the desired property. Here a two-dimensional linear autonomous system of Caputo FDEs with such a property will be constructed.

Since $\alpha \in (0, 1)$, the complex-valued Mittag-Leffler function $E_\alpha(z)$ has infinitely many zeros in $\mathbb{C}$ [2, Corollary 3.10, p. 30]. Take and fix $z^* \in \mathbb{C}$ such that $E_\alpha(z^*) = 0$. Let $\phi := \arg(z^*) \in (-\pi, \pi]$ and write $\lambda := \cos\phi + \iota \sin\phi$, where $\iota = \sqrt{-1} \in \mathbb{C}$.

Note that $E_\alpha(\overline{z^*}) = \overline{E_\alpha(z^*)} = 0$, since $\alpha \in \mathbb{R}$, where $\overline{w}$ denotes the complex conjugate a the complex number w. Moreover, $z^* \notin \mathbb{R}$ since $\alpha \in (0, 1)$. Hence $\lambda \notin \mathbb{R}$. Define

$$A := \begin{pmatrix} \cos\phi & \sin\phi \\ -\sin\phi & \cos\phi \end{pmatrix}.$$

Then A has two (complex) eigenvalues $\lambda, \overline{\lambda}$.

Consider the linear Caputo FDE

$$^C D_{0+}^\alpha x(t) = Ax(t), \quad t \geq 0. \tag{7.15}$$

This system has the asserted property. Indeed, by Diethelm [4, Theorem 7.15, p. 152], this linear Caputo FDE is explicitly solvable and has two following independent real solutions

$$x_1(t) := \begin{pmatrix} E_\alpha(\lambda t^\alpha) + E_\alpha(\bar\lambda t^\alpha) \\ \imath\left(E_\alpha(\lambda t^\alpha) - E_\alpha(\bar\lambda t^\alpha)\right) \end{pmatrix}, \quad x_2(t) := \begin{pmatrix} -\imath\left(E_\alpha(\lambda t^\alpha) - E_\alpha(\bar\lambda t^\alpha)\right) \\ E_\alpha(\lambda t^\alpha) + E_\alpha(\bar\lambda t^\alpha) \end{pmatrix}.$$

Write

$$u(t) := E_\alpha(\lambda t^\alpha) + E_\alpha(\bar\lambda t^\alpha) \quad \text{and} \quad v(t) := \imath\left(E_\alpha(\lambda t^\alpha) - E_\alpha(\bar\lambda t^\alpha)\right).$$

Then $u, v : \mathbb{R}^+ \to \mathbb{R}$ with $u(0) = 2$, $v(0) = 0$, and

$$x_1(t) = \begin{pmatrix} u(t) \\ v(t) \end{pmatrix}, \quad x_2(t) = \begin{pmatrix} -v(t) \\ u(t) \end{pmatrix}, \quad x_1(0) = \begin{pmatrix} 2 \\ 0 \end{pmatrix}, \quad x_2(0) = \begin{pmatrix} 0 \\ 2 \end{pmatrix}.$$

The general solution of (7.15) is

$$x(t) = ax_1(t) + bx_2(t) \tag{7.16}$$

where $a, b \in \mathbb{R}$ are arbitrary real constants. Let $T > 0$ be the unique finite positive number satisfying

$$\lambda T^\alpha = z^*.$$

Such a T exists uniquely due to the definitions of z^* and λ. Clearly $u(T) = v(T) = 0$, hence $x_1(T) = x_2(T) = (0, 0)^\mathrm{T}$. It follows from (7.16) that $x(T) = 0$ for any solution $x(t)$ of the Caputo FDE (7.15). $\qquad\square$

Remark 7.2 In fact, *all of the solutions* the Caputo FDE (7.15) equal $(0, 0)^\mathrm{T}$ at the time T, so they all intersect at this time.

    Theorem 7.5 reveals a distinguishing feature of Caputo FDEs in comparision with ODEs: different trajectories of an ODE cannot intersect, whereas those of Caputo FDEs may intersect.

7.4   Endnotes

This chapter is based primarily on Cong and Tuan [1] and Cong, Tuan and Trinh [3]. See also [5, 6]. In particular, Theorem 7.1 is from [1, Theorem 3.5].

References

1. Cong, N.D., Tuan, H.T.: Generation of nonlocal dynamical systems by fractional differential equations. J. Integral Equ. Appl. **29**, 585–608 (2017). https://doi.org/10.1216/JIE-2017-29-4-585
2. Gorenflo, R., Kilbas, A.A., Mainardi, F., Rogosin, S.V.: Mittag-Leffler Functions. Related Topics and Applications. Springer-Verlag, Berlin Heidelberg (2014)

3. Cong, N.D., Tuan, H.T., Hieu, T.: On asymptotic properties of solutions to fractional differential equations. J. Math. Anal. Appl. **484**, 123759 (2020)

4. Diethelm, K.: The Analysis of Fractional Differential Equations, Springer Lecture Notes in Mathematics, vol. 2004. Springer, Heidelberg (2010)

5. Diethelm, K.: On the separation of solutions of fractional differential equations. Fract. Calc. Appl. Anal. **11**(3), 259–268 (2008)

6. Diethelm, K., Ford, N.J.: Volterra integral equations and fractional calculus: do neighboring solutions intersect? J. Integral Equ. Appl. **24**(1), 25–37 (2012). https://doi.org/10.1216/JIE-2012-24-1-25

Chapter 8
Caputo Dynamical Systems

Abstract The first part of the chapter concerns the construction of nonlocal dynamical systems associated with scalar Caputo fractional differential equations and triangular systems. The key point is to establish the existence of a nonlocal two-parameter flow generated by their solutions. The second part deals with operator semi-groups associated with integral equations with degenerate kernels associated with initial value problems of Caputo fractional differential equations in finite-dimensional spaces.

Keywords Caputo fractional differential equations · Two-parameter flow · Evolution mapping · Nonlocal dynamical system · Semi-group of continuous operators

The solution $\phi(\cdot, x_0)$ of an autonomous ODE

$$\frac{dx}{dt} = g(x(t)), \ t > 0,$$

$$x(0) = x_0$$

on $\mathbb{R}^d$ generates a *semi-group* $\{\phi_t\}_{t \geq 0}$ of mappings on $\mathbb{R}^d$ defined by $\phi_t(x_0) := \varphi(t, x_0)$.

Definition 8.1 A family of mappings $\{\phi_t\}_{t \geq 0}$ on $\mathbb{R}^d$ is called a *semi-group* if

  (i)   the mapping $\phi_0(x_0) = x_0$ for all $x_0 \in \mathbb{R}^d$;
  (ii) $\phi_t \circ \phi_s = \phi_{t+s}$ for all $s, t \geq 0$ (semi-group property);
  (iii) $(t, x) \mapsto \phi_t(x)$ is continuous in both variables $t \geq 0$, $x \in \mathbb{R}^d$.

It is often called an autonomous *semi-dynamical system* on $\mathbb{R}^d$.

Note that the semi-group property is a consequence of the existence and uniqueness of solutions of the corresponding initial value problems. It means that solutions of an ODE to be concatenated.

In contrast the solutions of Caputo fractional differential equations depend not only on their initial value but also on all values up to the current time. This, in

   99
T. S. Doan et al., *Attractors of Caputo Fractional Differential Equations*,
SpringerBriefs in Mathematics, https://doi.org/10.1007/978-3-032-05511-8_8

particular, means that solutions cannot be concatenated and that Caputo FDEs on $\mathbb{R}^d$ cannot generate a semi-group on $\mathbb{R}^d$.

In Chap. 7 it was seen that any two trajectories of solutions of a one-dimensional Caputo FDE either coincide or do not intersect each other. This also holds for triangular systems of Caputo FDEs. It will be shown here that one-dimensional Caputo FDEs and triangular systems of Caputo FDEs generate *nonlocal fractional dynamical systems* on the appropriate space $\mathbb{R}^d$, $d \geq 1$.

In the higher dimensional case, two different trajectories of a Caputo FDE can intersect. Hence, higher dimensional Caputo FDEs do not, in general, generate such nonlocal dynamical system in the space $\mathbb{R}^d$. However, the corresponding Volterra integral equation does generate a semi-group in a suitable function space of continuous functions and this semi-group can be restricted to the solutions of the Caputo FDE.

8.1　Scalar Case

It will be shown here following Cong and Tuan [1] that one-dimensional Caputo FDEs generate a nonlocal two-parameter flow on $\mathbb{R}$.

Definition 8.2　A two-parameter family of mappings

$$\phi_{s,t}(\cdot) : \mathbb{R} \to \mathbb{R}, \qquad s, t \in [0, T],$$

is called *a two-parameter flow in* $\mathbb{R}$ if

  (i)　$(s, t, x) \mapsto \phi_{s,t}(x)$ is continuous in all three variables $s, t \in [0, T]$, $x \in \mathbb{R}$;

  (ii)　the mapping $\phi_{s,t}(\cdot)$ is a homeomorphism of $\mathbb{R}$ for any fixed $s, t \in [0, T]$;

 (iii)　this family satisfies the two-parameter flow property

$$\phi_{s,t} \circ \phi_{r,s} = \phi_{r,t} \qquad \text{for all} \quad r, s, t \in [0, T].$$

To apply this to Caputo FDEs requires the definition of the evolution mapping of the Caputo FDE (7.1):

$$^{C}D_{0+}^{\alpha} x(t) = g(x(t)), \quad t > 0,$$

where $\alpha \in (0, 1)$, $g : \mathbb{R} \to \mathbb{R}$ is Lipschitz continuous.

Definition 8.3　The mapping

$$\Phi_{0,\tau} : \mathbb{R} \to \mathbb{R}, \quad x_0 \mapsto \varphi(\tau, x_0), \tag{8.1}$$

is called *the evolution mapping* of Caputo FDE (7.1), where $x(\cdot, x_0)$ is the solution of the Caputo FDE (7.1) with the initial condition $\varphi(0, x_0) = x_0$ for an arbitrary $x_0 \in \mathbb{R}$ and $\varphi(\tau, x_0)$ is the evaluation of $\varphi(\cdot, x_0)$ at time $\tau \geq 0$.

Scalar Caputo FDEs generate two-parameter flows in $\mathbb{R}$.

Theorem 8.1 *The following statements hold for the one-dimensional Caputo FDE (7.1).*

1. *The evolution mapping $\Phi_{0,t}$ generated by the Caputo FDE (7.1) is a bijection for any $t \in [0, T]$.*
2. *The Caputo FDE (7.1) generates a two-parameter family of bijections on $[0, T]$ by its evolution mappings as follows*

$$\Phi_{s,t} := \Phi_{0,t} \circ \Phi_{0,s}^{-1} \quad \text{for all} \ \ s, t \in [0, T], \tag{8.2}$$

where $\Phi_{0,\cdot}$ is the evolution mapping of Caputo FDE (7.1) defined in Definition 8.3.
3. *The family $\Phi_{s,t}$, $s, t \in [0, T]$, generated by the Caputo FDE (7.1) is a two-parameter flow in $\mathbb{R}$.*
4. *If g is linear in x, then the two-parameter flow generated by the Caputo FDE (7.1) is a flow of linear operators.*

Proof (i) Fix $T_1 \in [0, T]$ arbitrarily. By Theorem 7.1, Φ_{0,T_1} is injective.

To show that Φ_{0,T_1} is surjective, it suffices to show that for an arbitrary $x^* \in \mathbb{R}$, the terminal value problem

$$^C D_{0+}^\alpha x(t) = g(x(t)), \tag{8.3}$$
$$x(T_1) = x^* \tag{8.4}$$

for $t \in [0, T_1]$, where g is continuous and satisfies the Lipschitz condition (7.2), has a continuous solution. Denote the solution of the Caputo FDE (8.3) satisfying the initial value condition $\hat{x}(0) = 0$ by $\hat{x}(\cdot)$.

Let L be the Lipschitz constant in the condition (7.2) and write

$$M_1 := \max_{0 \le t \le T_1} |\hat{x}(t)|, \qquad M_2 := |x^* - \hat{x}(T_1)| + M_1,$$

and

$$M_3 := \frac{M_2}{E_\alpha(-LT_1^\alpha)}, \qquad M_4 := M_1 + M_3 E_\alpha(LT_1^\alpha) + 1.$$

Clearly, $M_1 < M_2 < M_3 < M_4$. Define a function $\hat{g}$ on $\mathbb{R}$ as follows

$$\hat{g}(x) = \begin{cases} g(x), & \text{if} \ \ |x| \le M_4, \\ g\left(M_4 \frac{x}{|x|}\right), & \text{if} \ \ |x| > M_4, \end{cases} \tag{8.5}$$

and consider the terminal value problem of the FDE

$$^C D_{0+}^\alpha x(t) = \hat{g}(x(t)), \tag{8.6}$$
$$x(T_1) = x^*. \tag{8.7}$$

Since $\hat{g}$ is Lipschitz continuous with Lipschitz constant L and is bounded on $\mathbb{R}$, the terminal value problem (8.6)–(8.7) has at least one solution, say $x_1(\cdot)$ (see [2, Theorem 8]).

It remains to be shown that $x_1(\cdot)$ is the required solution of (8.3)–(8.4). To this end, notice that for all $t \in [0, T_1]$,

$$|\hat{x}(t)| \leq M_1 < M_4, \quad \text{and thus} \quad g(x(t)) = \hat{g}(x(t)), \quad t \in [0, T_1].$$

Therefore, $\hat{x}(\cdot)$ is the solution of the Caputo FDE (8.6) satisfying the initial value condition $x(0) = 0$. Applying Theorem 7.3 to the solutions $\hat{x}(\cdot)$ and $x_1(\cdot)$ of the Caputo FDE (8.6) gives the inequality

$$|\hat{x}(t) - x_1(t)| \geq |\hat{x}(0) - x_1(0)|E_\alpha\big(-Lt^\alpha\big) \geq |x_1(0)|E_\alpha(-LT_1^\alpha)$$

for any $t \in [0, T_1]$. Then, substituting $t = T_1$,

$$|\hat{x}(T_1) - x_1(T_1)| \geq |x_1(0)|E_\alpha(-LT_1^\alpha),$$

hence
$$|x_1(0)| \leq \frac{|x(T_1) - x^*|}{E_\alpha(-LT_1^\alpha)} \leq M_3.$$

Applying Theorem 7.4 to the solutions $\hat{x}(\cdot)$ and $x_1(\cdot)$ of the Caputo FDE (8.6) give the inequality
$$|x_1(t)| \leq |\hat{x}(t)| + |x_1(0)|E_\alpha(LT_1^\alpha) \leq M_4$$

which implies that $g(x(t)) = \hat{g}(x(t))$, $t \in [0, T_1]$. Thus $x_1(\cdot)$ is a solution of the Caputo FDE (8.3), and (i) is proved.

(ii) By (i), the evolution mapping of the Caputo FDE (7.1) are bijective, hence $\Phi_{s,t}$ is well defined by (8.2). The two-parameter flow property is easily verified from the definitions.

(iii) By (ii), the Caputo FDE (7.1) generates a two-parameter family of bijections $\Phi_{s,t}$ of $\mathbb{R}$ for all $s, t \in [0, T_1]$. From Theorems 7.3 and 7.4, it follows that the bijections $\Phi_{s,t}$ are homeomorphisms and Φ depends continuously in the three variables s, t, x.

(iv) Obvious. $\square$.

The two-parameter flow $\Phi_{s,t}$ in Theorem 8.1 was called the *nonlocal dynamical system* generated by Caputo FDE (7.1) in Cong and Tuan [1].

Remark 8.1 The two-parameter flow generated by the Caputo FDE (7.1) has memory. Although the past has an impact on the behaviour of the solutions, the solutions, nevertheless, still satisfy a two-parameter flow of homeomorphisms by the construction here. The flow is, in general, α-Hölder continuous, but not C^1. It differs from the two-parameter semi-group generated by the solutions of a nonautonomous ODE, see [3], which is local, i.e., the solutions depend only on the current value of the state and not on their history.

8.2 Triangular Systems of Caputo FDEs

Now consider the d-dimensional triangular system of Caputo FDEs (8.8):

$$
\begin{cases}
{}^{C}D_{0+}^{\alpha}x_1(t) & = g_1(x_1(t)), \\
{}^{C}D_{0+}^{\alpha}x_2(t) & = g_2(x_1(t), x_2(t)), \\
\quad \cdots \\
{}^{C}D_{0+}^{\alpha}x_d(t) & = g_d(x_1(t), x_2(t), \cdots, x_d(t)),
\end{cases}
\tag{8.8}
$$

where $x(\cdot) = \big(x_1(\cdot), \ldots, x_d(\cdot)\big)^{\mathrm{T}} \in \mathbb{R}^d$ and $g = (g_1, \ldots, g_d)^{\mathrm{T}}$ is globally Lipschitz continuous.

Similarly to the one-dimensional case, for any $\tau \in [0, T]$ the mapping

$$
\Phi_{0,\tau} : \mathbb{R}^d \to \mathbb{R}^d, \quad x_0 \mapsto \varphi(\tau, x_0)
$$

called *the evolution mapping of* (8.8), where $\varphi(\cdot, x_0)$ is the solution of (8.8) starting from and arbitrary initial value $x(0) = x_0$ and $\varphi(\tau, x_0)$ is the evaluation of $\varphi(\cdot, x_0)$ at τ.

It can be shown that by the same arguments as in the previous section that the evolution mapping $\Phi_{0,t}$ of the triangular FDE (8.8) is a bijection for any $t \in [0, T]$ and generates a two-parameter flow in $\mathbb{R}^d$.

Theorem 8.2 *The triangular Caputo FDE (8.8) generates a two-parameter flow in* $\mathbb{R}^d$ *defined by*

$$
\phi_{s,t} := \Phi_{0,t} \circ \Phi_{0,s}^{-1} \quad \text{for all} \;\; s, t \in [0, T],
$$

where $\Phi_{0,t}, t \in [0, T]$, *is the evolution mapping of the triangular Caputo FDE (8.8).*

Proof The proof is similar to that of Theorem 8.1. $\qquad\square$.

8.3 General Case: Caputo Semi-Group

The counterexample in Theorem 7.5 shows that a Caputo FDE of dimension $d \geq 2$ does not, in general, generate a two-parameter flow on $\mathbb{R}^d$. Hence it does not, in general, generate a nonlocal dynamical system on $\mathbb{R}^d$.

Nevertheless, Doan and Kloeden [4] have shown that a general autonomous Caputo fractional differential equation

$$
{}^{C}D_{0+}^{\alpha}x(t) = g(x(t)), \quad \text{where } g : \mathbb{R}^d \to \mathbb{R}^d,
\tag{8.9}
$$

generates a semi-dynamical system on the function space $\mathfrak{C}$ of continuous functions $f : [0, \infty) \to \mathbb{R}^d$ with the topology uniform convergence on compact subsets. This topology is induced by the metric

$$\rho(f,h) := \sum_{n=1}^{\infty} \frac{1}{2^n} \rho_n(f,h), \quad \text{where } \rho_n(f,h) := \frac{\sup_{t\in[0,n]} \|f(t) - h(t)\|}{1 + \sup_{t\in[0,n]} \|f(t) - h(t)\|}.$$

Define the operators $T_\tau : \mathfrak{C} \to \mathfrak{C}$, $\tau \geq 0$, by

$$(T_\tau f)(\theta) = f(\tau + \theta) + \frac{1}{\Gamma(\alpha)} \int_0^\tau (\tau + \theta - s)^{\alpha-1} g(x_f(s)) \, ds, \qquad \theta \geq 0, \quad (8.10)$$

where x_f is a solution of the singular Volterra integral equation

$$x_f(t) = f(t) + \frac{1}{\Gamma(\alpha)} \int_0^t (t - s)^{\alpha-1} g(x_f(s)) \, ds. \qquad (8.11)$$

The solutions of the Caputo FDE corresponds to the case that $f = id_{x_0}$, i.e., $f(t) \equiv id_{x_0}$ for all $t \geq 0$.

For notational convenience in the proof, the kernel $\frac{(t-s)^{\alpha-1}}{\Gamma(\alpha)}$ in the above equations will be denoted by $a(t, s)$ and the equations written more compactly as

$$(T_\tau f)(\theta) = f(\tau + \theta) + \int_0^\tau a(\tau + \theta, s) g(x_f(s)) \, ds, \qquad \theta \geq 0,$$

and

$$x_f(t) = f(t) + \int_0^t a(t, s) g(x_f(s)) \, ds.$$

Theorem 8.3 *Suppose that the vector field g is globally Lipschitz continuous. The integral equation (8.11) generalisation of the autonomous Caputo FDE (8.9) generates a semi-group of continuous operators $\{T_\tau, \tau \geq 0\}$ on the space $\mathfrak{C}$.*

Proof First show that $T_\tau : C \to C$ is continuous. Let $f, h \in C$. Then, by (8.10)

$$\|T_\tau f(\theta) - T_\tau h(\theta)\| \leq \|f(\tau + \theta) - h(\tau + \theta)\|$$
$$+ L \sup_{s\in[0,\tau]} \|x_f(s) - x_h(s)\| \int_0^\tau a(\tau + \theta, s) \, ds,$$

where L is the Lipschitz constant of the vector field g. A direct computation yields that

$$\int_0^\tau a(\tau + \theta, s) \, ds = \frac{1}{\Gamma(\alpha)} \int_0^\tau (\tau + \theta - s)^{\alpha-1} \, ds = \frac{1}{\alpha \, \Gamma(\alpha)} \left((\tau + \theta)^\alpha - \theta^\alpha \right).$$

Now, choose and fix $k \in \mathbb{N}$ with $k \geq \tau$. Then,

$$\sup_{\theta\in[0,n]} \|T_\tau f(\theta) - T_\tau g(\theta)\| \leq \sup_{t\in[0,k+n]} \|f(t) - h(t)\| + \frac{L(k+n)^\alpha}{\alpha\Gamma(\alpha)} \sup_{s\in[0,\tau]} \|x_f(s) - x_h(s)\|.$$

Using inequality $\frac{x}{1+x} \leq \frac{y}{1+y} + z$ provided that x, y, z are non-negative and $x \leq y + z$ yields that

$$
\rho_n(T_\tau f, T_\tau h) \leq \rho_{n+k}(f, h) + \frac{L(k+n)^\alpha}{\alpha \Gamma(\alpha)} \sup_{s \in [0,\tau]} \|x_f(s) - x_h(s)\|.
$$

Thus,

$$
\rho(T_\tau f, T_\tau h) \leq 2^k \rho(f, h) + \frac{Lc}{\alpha \Gamma(\alpha)} \sup_{s \in [0,\tau]} \|x_f(s) - x_h(s)\|,
$$

where $c := \sum_{n=1}^{\infty} 2^{-n}(k+n)^\alpha$. By Lemma 3.4, $\sup_{s \in [0,\tau]} \|x_f(s) - x_h(s)\| \to 0$ as $\rho(f, h) \to 0$. Consequently, $T_\tau(\cdot)$ is continuous.

To complete the proof, it needs to be shown that $\{T_\tau : \tau \geq 0\}$ forms a semi-group. Note that

$$
x_f(t) = f(t) + \int_0^t a(t, s)g(x_f(s))\, ds.
$$

Then

$$
\begin{aligned}
x_f(t + \tau) &= f(t + \tau) + \int_0^{t+\tau} a(t + \tau, s)g(x_f(s))\, ds \\
&= f(t + \tau) + \left(\int_0^\tau + \int_\tau^{t+\tau} \right) a(t + \tau, s)g(x_f(s))\, ds \\
&= (T_\tau f)(t) + \int_\tau^{t+\tau} a(t + \tau, s)g(x_f(s))\, ds \\
&= (T_\tau f)(t) + \int_0^t a(t + \tau, r + \tau)g(x_f(r + \tau))\, dr, \qquad (r = s - \tau), \\
&= (T_\tau f)(t) + \int_0^t a(t, r)g(x_f(r + \tau))\, dr.
\end{aligned}
$$

Hence by the existence and uniqueness of solutions $x_f(t + \tau) = \psi(t)$, where $\psi(t)$ is a solution of

$$
\psi(t) = (T_\tau f)(t) + \int_0^t a(t, s)g(\psi(s))\, ds.
$$

Moreover,

$$
\begin{aligned}
(T_\sigma (T_\tau f)))\, (\theta) &= (T_\tau f)(\sigma + \theta) + \int_0^\sigma a(\sigma + \theta, s)g(\psi(s))\, ds \\
&= f(\tau + \sigma + \theta) + \int_0^\tau a(\tau + \sigma + \theta, s)g(x_f(s))\, ds \\
&\quad + \int_0^\sigma a(\sigma + \theta, s)g(\psi(s))\, ds
\end{aligned}
$$

$$= f(\tau + \sigma + \theta) + \int_0^\tau a(\tau + \sigma + \theta, s) g(x_f(s))\, ds$$

$$+ \int_\tau^{\tau+\sigma} a(\sigma + \theta, r - \tau) g(\psi(r - \tau))\, dr, \quad (r = s + \tau).$$

Since $a(\sigma + \theta, r - \tau) = a(\tau + \sigma + \theta, r)$ and $\psi(r - \tau) = x_f(r)$ it follows that

$$(T_\sigma\,(T_\tau f)))\,(\theta) = f(\tau + \sigma + \theta) + \int_0^\tau a(\tau + \sigma + \theta, s) g(x_f(s))\, ds$$

$$+ \int_\tau^{\tau+\sigma} a(\tau + \sigma + \theta, r) g(x_f(r))\, dr.$$

This gives

$$(T_\sigma\,(T_\tau f))\,(\theta) = f(\tau + \sigma + \theta) + \int_0^{\tau+\sigma} a(\tau + \sigma + \theta, s) g(x_f(s))\, ds.$$

On the other hand from the definition of the operator as in (8.10)

$$(T_{\sigma+\tau} f)\,(\theta) = f(\tau + \sigma + \theta) + \int_0^{\tau+\sigma} a(\tau + \sigma + \theta, s) g(x_f(s))\, ds.$$

This means that

$$(T_{\sigma+\tau} f)\,(\theta) = (T_\sigma\,(T_\tau f))\,(\theta), \qquad \forall \tau, \theta, \sigma \geq 0,\ f \in \mathfrak{C},$$

that is

$$T_{\sigma+\tau} f = T_\sigma\,(T_\tau)\, f, \qquad \forall \tau, \sigma \geq 0,\ f \in \mathfrak{C}. \qquad \square.$$

Remark 8.2 The above result also hold for autonomous Caputo fractional differential equations with a substantial time derivative, i.e., of the form

$$x(t) = x_0 + \frac{1}{\Gamma(\alpha)} \int_0^t (t - s)^{\alpha-1} e^{-\beta(t-s)} g(x(s))\, ds,$$

where $\beta > 0$. This can be seen by replacing $a(t, s)$ by

$$\tilde{a}(t, s) := \frac{1}{\Gamma(\alpha)} (t - s)^{\alpha-1} e^{-\beta(t-s)}, \qquad 0 \leq s < t.$$

Note that $0 < \tilde{a}(t, s) \leq a(t, s)$.

8.4 Endnotes

The nonlocal two-parameter flow results are taken from Cong and Tuan [1]. For two-parameter semi-groups generated by nonautonomous ODEs see [3, 5].

The proof of Theorem 8.3 is taken from Doan and Kloeden [4]. It follows closely Chap. XI, pages 178–179, in Sell [6], where a vector field g in the integral equation (11.5) is called *admissible* if it has a globally defined unique solution for each $f \in \mathfrak{C}$ with continuity in initial data. This holds if g is globally Lipschitz continuous as above, but weaker assumptions are also possible. For example g is admissble if it satisfies a local Lipschitz condition and a dissipativity condition (1.2). See also Cong [7].

References

1. Cong, N.D., Tuan, H.T.: Generation of nonlocal dynamical systems by fractional differential equations. J. Integral Equ. Appl. **29**, 585–608 (2017). https://doi.org/10.1216/JIE-2017-29-4-585
2. Benchohra, M., Hamani, S., Ntouyas, S.K.: Boundary value problems for differential equations with fractional order. Surveys in Mathematics and Its Applications **3**, 1–12 (2008)
3. Kloeden, P.E., Yang, M.: Introduction to Nonautonomous Dynamical Systems and their Attractors. World Scientific Publishing Co., Inc, Singapore (2021)
4. Doan, T.S., Kloeden, P.E.: Semi-dynamical systems generated by autonomous Caputo fractional differential equations. Vietnam J. Math. **49**, 1305–1315 (2021). https://doi.org/10.1007/s10013-020-00464-6
5. Kloeden, P.E., Rasmussen, M.: Nonautonomous dynamical systems. American Mathematical Society, Providence (2011)
6. Sell, G.R.: Topological dynamics and ordinary differential equations. Van Nostrand Reinhold Mathematical Studies, London (1971)
7. Cong, N.D.: Semigroup property of fractional differential operators and its applications. Discr. Continuous Dyn. Syst. Ser. B **28**, 1–12 (2023)

Chapter 9
Dissipative Caputo FDEs

Abstract Qualitative properties of solutions to dissipative Caputo fractional differential equations are investigated. The existence and uniqueness of solutions are established and then the existence of absorbing sets for solutions. The omega-limit sets of the nonlocal two-parameter flow are analyzed.

Keywords Dissipative Caputo fractional differential equations · Existence and uniqueness of solutions · Attractor · Omega limit set · Two-parameter flow

Suppose that the vector field $g : \mathbb{R}^d \to \mathbb{R}^d$ of the Caputo FDE of order $\alpha \in (0, 1)$

$$^{C}D_{0+}^{\alpha} x(t) = g(x(t)), \quad t > 0 \tag{9.1}$$

on $\mathbb{R}^d$ is locally Lipschitz continuous, i.e.,

Assumption 9.1 A function $g : \mathbb{R}^d \to \mathbb{R}^d$ is said to be locally Lipschitz if for every $R > 0$ there exists $L_R > 0$ such that

$$\|g(x) - g(y)\| \leq L_R \|x - y\| \quad \text{for all} \quad \|x\|, \|y\| \leq R. \tag{9.2}$$

Suppose also that the vector field g satisfies a dissipativity condition such as

$$\left\langle x, g(x) \right\rangle \leq a - b\|x\|^2, \tag{9.3}$$

where $a, b > 0$ and $\left\langle \cdot, \cdot \right\rangle$ is the scalar product in $\mathbb{R}^d$.

Then, by the Tuan-Trinh inequality (2.1), the solutions of the Caputo FDE (9.1) satisfy

$$^{C}D_{0+}^{\alpha} \|x(t)\|^2 \leq \left\langle x(t), {}^{C}D_{0+}^{\alpha} x(t) \right\rangle = 2\left\langle x(t), g(x(t)) \right\rangle \leq 2a - 2b\|x(t)\|^2, \quad t > 0. \tag{9.4}$$

T. S. Doan et al., *Attractors of Caputo Fractional Differential Equations*,
SpringerBriefs in Mathematics, https://doi.org/10.1007/978-3-032-05511-8_9

It follows then by Lemma 2.3 that

$$\|x(t)\|^2 \leq \|x_0\|^2 E_\alpha(-2bt^\alpha) + \frac{a}{b}\left(1 - E_\alpha(-2bt^\alpha)\right), \quad t \geq 0. \tag{9.5}$$

9.1 Existence: Local Lipschitz Case with Dissipativity

For many dissipative Caputo FDEs the vector fields are often only continuously differentiable and hence only locally Lipschitz rather than globally Lipschitz.

Nevertheless, the dissipativity condition (9.3) with a local Lipschitz condition suffices to ensure global existence and uniqueness of solutions.

Theorem 9.1 *Assume that the vector field g satisfies a local Lipschitz condition and the dissipativity condition (9.3). Then, for any $T > 0$ and for each initial value $x_0 \in \mathbb{R}^d$ the Caputo FDE (9.1) has a unique solution $\varphi(\cdot, x_0)$ on the interval $[0, T]$.*

Proof Consider a large ball $B_R = \{x \in \mathbb{R}^d \; : \; \|x\| \leq R\}$ with $R^2 > \frac{a}{b}$ and define

$$g_R(x) := \begin{cases} g(x), & \|x\| \leq R, \\ g(xR/\|x\|), & \|x\| \geq R. \end{cases}$$

This function satisfies a global Lipschitz condition, so the existence and uniqueness Theorem 3.1 holds for the Caputo FDE

$$^{C}D_{0+}^\alpha x = g_R(x).$$

Hence its solutions exist globally and are unique for all initial values.

Moreover, g_R satisfies the dissipativity condition (9.3) inside the ball B_R, so the inequality (9.4) with $w(t) = \|x(t)\|^2$ can be used provided the solutions starting in this ball remain in it.

Suppose $\|x(0)\| < R$ and let $\tau > 0$ be the first time (if it exists) for which the solution equals R, i.e., $\|x(t)\| < \|x(\tau)\| = R$ for $0 \leq t < \tau$. This corresponds to the solution of the original Caputo FDE, so the inequality (9.5) can be applied to give

$$R^2 = \|x(\tau)\|^2 \leq \|x(0)\|^2 E_\alpha(-2b\tau^\alpha) + \frac{a}{b}\left(1 - E_\alpha(-2b\tau^\alpha)\right)$$

$$< R^2 E_\alpha(-2b\tau^\alpha) + R^2\left(1 - E_\alpha(-2b\tau^\alpha)\right) = R^2.$$

This is a contradiction, so $\|x(t)\| < R$ for all $t > 0$.

Since the choice of $R^2 > \frac{a}{b}$ was otherwise arbitrary, the result holds for all such R and, hence, for all initial values $x_0 \in \mathbb{R}^d$. $\qquad\square$

9.2 Absorbing Sets

It also follows from the dissipativity property (9.3) that the Caputo FDE (9.1) on $\mathbb{R}^d$ has an absorbing set, i.e., a closed and bounded subset $\mathcal{B}$ of $\mathbb{R}^d$ which all solutions enter within a finite time. In fact, it absorbs solutions starting in bounded subsets of initial conditions uniformly in time.

Theorem 9.2 *Suppose that the vector field g satisfies the dissipativity condition (9.3). Then the solutions of the Caputo FDE (9.1) are absorbed by the subset*

$$\mathcal{B} := \left\{ x \in \mathbb{R}^d \ : \ \|x\|^2 \leq 1 + \frac{a}{b} \right\}.$$

Moreover, this set is positively invariant.

Proof Consider initial values $x_0 \in D$, an arbitrary bounded subset of $\mathbb{R}^d$. There exists a $T(D) \geq 0$ such that

$$\|x\|^2 E_\alpha(-2bt^\alpha) \leq 1 \quad \text{for } t \geq T(D) \ \text{ and } \ x \in D.$$

Hence, by inequality (9.5),

$$\|x(t)\|^2 \leq 1 + \frac{a}{b}, \qquad t \geq T(D).$$

This shows that the set $\mathcal{B}$ is an absorbing set.

Moreover, for $\|x_0\|^2 \leq 1 + \frac{a}{b}$, (9.5) gives

$$\|x(t)\|^2 = \|x_0\|^2 E_\alpha(-2bt^\alpha) + \frac{a}{b}\left(1 - E_\alpha(-2bt^\alpha)\right)$$

$$\leq 1 + \frac{a}{b}, \quad t \geq 0.$$

This establishes positive invariance of $\mathcal{B}$ and completes the proof. $\qquad\square$

Thus every solution of the dissipative Caputo FDE (9.1) entering into $\mathcal{B}$ stays there, as does any solution starting in $\mathcal{B}$, i.e., $\mathcal{B}$ is positive invariant. The importance of the absorbing set is that all future, limiting dynamics occurs in it.

9.2.1 Attractors and Omega Limit Sets for ODEs

The set $\mathcal{B}$ defined in Theorem 9.2 is also a positive invariant absorbing set for the corresponding autonomous ODE in $\mathbb{R}^d$ with the same autonomous vector field $g(x)$. Its solution mapping $\varphi(t, x_0)$ generates an autonomous semi-dynamical system, i.e., satisfies a semi-group property, and has a global attractor $\mathcal{A}$, i.e.,

Definition 9.1 An attractor $\mathcal{A}$ of an autonomous semi-dynamical system φ on $\mathbb{R}^d$

(i)   is a nonempty compact subset of $\mathbb{R}^d$;
(ii) is invariant, ie., $\mathcal{A} = \varphi(t, \mathcal{A})$ for all $t \geq 0$;
(iii) attracts bounded subsets D of $\mathbb{R}^d$, i.e., $\lim_{t \to \infty} \mathrm{dist}_{\mathbb{R}^d}(\varphi(t, D), \mathcal{A}) = 0$.

Here $\varphi(t, D) = \cup_{x_0 \in D} \varphi(t, x_0)$ and $\mathrm{dist}_{\mathbb{R}^d}(A, B)$ denotes the Hausdorff semi-distance between two subsets A and B of $\mathbb{R}$, i.e.,

$$\mathrm{dist}_{\mathbb{R}^d}(A, B) := \sup_{a \in A} \mathrm{dist}_{\mathbb{R}^d}(a, B), \quad \mathrm{dist}_{\mathbb{R}^d}(a, B) := \inf_{b \in B} |a - b|.$$

It can be shown that global attractor is given by

$$\mathcal{A} = \bigcap_{t \geq 0} \varphi(t, \mathcal{B}) = \Omega_{\mathcal{B}}, \tag{9.6}$$

where $\Omega_{\mathcal{B}}$ is the omega limit set defined by

$$\Omega_{\mathcal{B}} := \left\{ y \in \mathcal{B} : \exists\, x_{0,n} \in \mathcal{B}, t_n \to \infty \text{ such that } \varphi(t_n, x_{0,n}) \to y \right\}.$$

The global attractor consist of all bounded entire solutions of the ODE, i.e., those defined for all $t \in \mathbb{R}^d$ and, in particular, the steady state solutions. In the ODE case it suffices to consider solutions starting inside $\mathcal{B}$ since all solutions starting outside of $\mathcal{B}$ enter into it in finite time and can be concatenated with a solution starting inside it.

The theory of autonomous semi-dynamical systems holds more generally on Banach state spaces with suitable modifications. See [1–3]. It implies the existence of a global attractor under appropriate assumptions for semi-dynamical systems on a Banach space.

Theorem 9.3 *Suppose that the semi-dynamical system $\{\varphi_t, t \in \mathbb{R}^+\}$ on a Banach space X has a closed and bounded positively invariant absorbing set $\mathcal{B}$ in X and is asymptotically compact. Then the semi-dynamical system $\{\varphi_t, t \in \mathbb{R}^+\}$ has a global attractor given by*

$$\mathcal{A} = \bigcap_{t \geq 0} \varphi_t(\mathcal{B}).$$

The proof is considerably simpler for ODEs in $\mathbb{R}^d$ since closed and bounded subsets are compact and continuous mappings take compact sets to compact sets.

There are many generalizations [1–3] of Theorem 9.3 to "universes" of subsets which are attracted, such as compact sets or bounded sets or just individual points.

9.2.2 Omega Limit Sets of Caputo FDEs

The solutions of a Caputo FDE do not generate a semi-group on $\mathbb{R}^d$. In particular, this means they cannot have an attractor in $\mathbb{R}^d$ as defined by (9.6).

However, omega limit points still provide relevant information about future asymptotic behaviour in $\mathbb{R}^d$. Since, in general, the solutions of Caputo FDE (9.1) cannot be concatenated, there may be omega limit points of solutions starting outside the absorbing set $\mathcal{B}$ that cannot be concatenated with solutions starting inside $\mathcal{B}$; there may be additional omega limit points that are not in $\Omega_{\mathcal{B}}$. It is thus necessary to consider the set of all possible omega limit points

$$\Omega^* = \overline{\bigcup \left\{ y \in \mathbb{R}^d \ : \ \exists \{x_{0,n}\}_{n \in \mathbb{Z}}, \text{bnd'd}, \ t_n \to \infty \text{ such that } \varphi(t_n, x_{0,n}) \to y \right\}} \subset \mathcal{B}. \tag{9.7}$$

Note that $\Omega^* \subset \mathcal{B}$, since $\mathcal{B}$ is an absorbing set.

Strictly speaking, Ω^* not an attractor in the above sense, but it is an attracting set containing all the limiting dynamics of the Caputo FDE in $\mathbb{R}^d$. Nevertheless, characterises and, in fact, determines the attractor of the general Caputo semi-dynamical system on the function space $\mathfrak{C}$, when it exists, see Chap. 11.

9.3 Omega Limit Sets for Caputo Nonlocal Two-Parameter Flows

This section considers the asymptotic behaviour of nonlocal two-parameter flows generated by dissipative Caputo FDEs, in particular their omega limit sets.

In Chap. 8 it was shown that Caputo FDEs (9.1) with a scalar or triangular vector field do generate a nonlocal two-parameter flow $\{\phi_{s,t}\}$, $0 \leq s \leq t$, on $\mathbb{R}^d$ via the evolution mapping $\Phi_{0,t}(x_0) := \varphi(t, x_0)$, i.e., with $\phi_{s,t}(x_0) := \Phi_{0,t} \circ \Phi_{0,s}^{-1}(x_0)$ for $0 \leq s \leq t$. In particular, the two-parameter semi-group property holds, $\phi_{s,t} := \phi_{s,t} \circ \phi_{0,s}$, and the mappings $\phi_{s,t}$ are homeomorphisms.

A compact set $\mathcal{B}$ in $\mathbb{R}^d$ is called *a compact absorbing set* of a nonlocal two-parameter flow ϕ if it absorbs every bounded set D starting from 0, i.e., there exists a time $T_D > 0$ such that

$$\phi_{0,t}(D) \subset \mathcal{B}, \quad t \geq T_D.$$

Note that by the homeomorphic property such an absorbing set $\mathcal{B}$ absorb every bounded set D starting from any $\tau > 0$ also. Indeed, since $\phi_{0,\tau}^{-1}(D)$ remains bounded,

$$\phi_{\tau,t}(D) = \phi_{0,t}\left(\phi_{0,\tau}^{-1}(D)\right) \subset \mathcal{B}, \quad t \geq T_{\phi_{0,\tau}^{-1}(D)}.$$

For a bounded subset D of $\mathbb{R}^d$ and $\tau \geq 0$, define

$$\Omega_{D,\tau} := \left\{ \bar{x} \in \mathcal{B} \ : \ \exists x_{0,n} \in D, t_n \to \infty \text{ such that } \phi_{\tau,t_n}(x_{0,n}) \to \bar{x} \right\}.$$

Lemma 9.1 $\Omega_{D,\tau}$ *is compact for every* $\tau \geq 0$ *and bounded subset* D *of* $\mathbb{R}^d$.

Proof Since $\Omega_{D,\tau}$ is contained in the compact set $\mathcal{B}$ it suffices to show that it is a closed set. Let $\bar{x}_n \in \Omega_{D,\tau}$ with $\bar{x}_n \to \bar{x}^*$ as $n \to \infty$, i.e., given $\varepsilon > 0$ here exists $N(\varepsilon) \in \mathbb{N}$ such that

$$\|\bar{x}_n - \bar{x}^*\| < \varepsilon/2, \quad n \geq N(\varepsilon).$$

For each $n \in \mathbb{N}$ there exist $x_j^{(n)} \in D$ and $t_j^{(n)} \to \infty$ as $j \to \infty$ such that $\phi_{\tau,t_j^{(n)}}(x_j^{(n)}) \to \bar{x}_n$ as $j \to \infty$, i.e., there exists $J_n(\varepsilon) \in \mathbb{N}$ such that

$$\|\phi_{\tau,t_j^{(n)}}(x_j^{(n)}) - \bar{x}_n\| < \varepsilon/2, \quad j \geq J_n(\varepsilon).$$

Let $\tau^{(0)} = \tau$ and $j_0^*(\varepsilon) = 1$. Then for $n \geq 1$ pick $j_n^*(\varepsilon) \geq J_n(\varepsilon)$ such that

$$\tau^{(n)} = t_{j_n^*(\varepsilon)}^{(n)} < 1 + t_{j_n^*(\varepsilon)}^{(n)} < t_{j_{n+1}^*(\varepsilon)}^{(n+1)} = \tau^{(n+1)},$$

so $\tau^{(n)} \to \infty$ as $n \to \infty$. Define $z_n = x_{J_n(\varepsilon)}^{(n)}$. Then

$$\|\phi_{\tau,\tau^{(n)}}(z_n) - \bar{x}_n\| = \|\phi_{\tau,t_{j_n^*(\varepsilon)}^{(n)}}(x_{J_n(\varepsilon)}^{(n)}) - \bar{x}_n\| < \varepsilon/2, \quad n \geq N(\varepsilon),$$

so

$$\|\phi_{\tau,\tau^{(n)}}(z_n) - \bar{x}^*\| < \varepsilon, \quad n \geq N(\varepsilon),$$

i.e., $\bar{x}^* \in \Omega_{D,\tau}$. Thus $\Omega_{D,\tau}$ is a closed subset of the compact set $\mathcal{B}$. $\qquad\square$

Now define

$$\Omega_\tau^* = \overline{\bigcup \{\Omega_{D,\tau} \; : \; \text{all bnd'd } D \subset \mathbb{R}^d\}}, \quad \tau \geq 0.$$

Obviously, $\Omega_\tau^* \subset \mathcal{B}$ is also compact. It easily shown that Ω_τ^* is the minimal compact set which is forward attracting starting from at τ, i.e., for any bounded set D in $\mathbb{R}^d$,

$$\lim_{t \to \infty} \mathrm{dist}_{\mathbb{R}^d}\left(\phi_{\tau,t}(D), \; \Omega_\tau^*\right) = 0.$$

The next result follows from the particular definition of the Caputo nonlocal two-parameter flow.

Lemma 9.2 $\Omega_\tau^* = \Omega_0^*$ *for all* $\tau \geq 0$.

Proof Let $\bar{x} \in \Omega_{D,0} \subset \Omega_0^*$, so there exist $x_{0,n} \in D$ and $t_n \to \infty$ such that $\phi_{0,t_n}(x_{0,n}) \to \bar{x}$.

Define $y_{0,n} := \phi_{0,\tau}(x_{0,n}) \subset \phi_{0,\tau}(D)$. Then $x_{0,n} := \phi_{0,\tau}^{-1}(y_{0,n})$ and

$$\phi_{\tau,t_n}(y_{0,n}) = \phi_{\tau,t_n} \circ \phi_{0,\tau}(x_{0,n}) = \phi_{0,t_n}(x_{0,n}) \to \bar{x}.$$

Hence $\bar{x} \in \Omega_{\phi_{0,\tau}(D),\tau} \subset \Omega_\tau^*$. Since $\bar{x} \in \Omega_{D,0} \subset \Omega_0^*$ was otherwise arbitrary, this means that $\Omega_0^* \subset \Omega_\tau^*$.

Now let $\bar{y} \in \mathbb{R}^d$ and define $y_{0,n} \in \varphi(\tau, D)$ for some bounded subset D and suppose that $\phi_{\tau,t_n}(y_{0,n}) \to \bar{y}$, hence $\bar{y} \in \Omega_{\phi_{0,\tau}(D),\tau}$. Take $x_{0,n} = \phi_{0,\tau}^{-1}(y_{0,n}) \in D$, then

$$\phi_{0,t_n}(x_{0,n}) = \phi_{0,t_n}(x_{0,n}) = \phi_{0,t_n} \circ \phi_{0,\tau}^{-1}(y_{0,n}) = \phi_{\tau,t_n}(y_{0,n}) \to \bar{y},$$

so $\bar{y} \in \Omega_{D,0} \subset \Omega_0^*$. This implies $\Omega_\tau^* \subset \Omega_0^*$. □

The subscript on Ω_0^* can thus be omitted.

Lemma 9.3 $\Omega^* = \Omega_{\mathcal{B}}$.

Proof It is clear that $\Omega_{\mathcal{B}} \subset \Omega^*$.

Let D be a bounded subset of $\mathbb{R}^d$. Then there is a T_D such that $\widehat{D} := \varphi(T_D, D) \subset \mathcal{B}$ and hence $\Omega_{\widehat{D}} \subset \Omega_{\mathcal{B}}$. But

$$\phi_{T_D,t}(\widehat{D}) = \phi_{0,t} \circ \phi_{0,T_D}^{-1}(\widehat{D}) = \phi_{0,t}(D), \quad t \geq T_D.$$

This implies that $\Omega_D = \Omega_{\widehat{D},T_D} \subset \Omega_{\mathcal{B}}$. But $\Omega_D \subset \Omega^*$ and D was arbitrary, so $\Omega^* \subset \Omega_{\mathcal{B}}$. □

This means that Ω_0^* is the common minimal compact set which is forward attracting no matter at what time the initial values were taken. Cui et al. [4] called this set the *minimal compact attractor* of ϕ, denoted by $\mathcal{A}_{\min}$. More precisely, it follows that $\mathcal{A}_{\min} := \Omega_0^*$ is the minimal compact set satisfying

$$\lim_{t \to \infty} \mathrm{dist}_{\mathbb{R}^d}\left(\phi_{\tau,t}(D),\ \mathcal{A}_{\min}\right) = 0$$

for all bounded set $D \subset \mathbb{R}^d$ and $\tau \geq 0$. By [4, Theorem 3.18], then

Lemma 9.4 $\mathcal{A}_{\min} = \Omega_{\mathcal{B}}$.

Thus for the Caputo nonlocal two-parameter flow, when it exists, all of the limiting dynamics is contained in the set $\Omega_{\mathcal{B}}$. It also contains all of the limit points of solutions starting outside of $\Omega_{\mathcal{B}}$ by the definition of the nonlocal two-parameter flow and not by the concatenation of solutions.

Note that $\mathcal{A}_{\min}$ is not assumed to be invariant [4]. In Theorem 10.1 it will be seen that in the scalar case $\Omega_{\mathcal{B}}$ consists of steady state solutions and heteroclinic trajectories joining them, so is invariant under $\phi_{0,t}$. A similar situtation also holds for Caputo FDEs with certain kinds of triangular vector fields, see Theorem 10.3.

9.4　Endnotes

Theorems 9.1 and 9.2 are taken from [5].

See [1–3] for autonomous semi-dynamical systems and their attractors and omega limit sets. There are many generalizations of Theorem 9.3 to "universes" of subsets which are attracted, such as compact sets or bounded sets or just individual points.

The results in Sect. 9.3 are new. Cui et al. [4] investigated minimal attractors of more general nonautonomous dynamical systems in Banach spaces, where they are assumed to be closed and bounded rather than compact.

References

1. Hale, J.K., Koçak, H.: Dynamics and Bifurcations. Springer-Verlag, New York (1991)
2. Kloeden, P.E., Rasmussen, M.: Nonautonomous Dynamical Systems. American Mathematical Society, Providence (2011)
3. Kloeden, P.E., Yang, M.: Introduction to nonautonomous dynamical systems and their attractors. World Scientific Publishing Co., Inc, Singapore (2021)
4. Hongyong, C., Figueroa-Lopez, R.N., Langa, J.A., Nascimento, M.J.D.: Forward attraction of nonautonomous dynamical systems and applications to Navier-Stokes equations. SIAM J. Appl. Dyn. Syst. **23** (2024). https://doi.org/10.1137/23M1626384
5. Kloeden, P.E.: An elementary inequality for dissipative Caputo fractional differential equations. Fract. Calc. Appl. Anal. **26**, 2166–2174 (2023). https://doi.org/10.1007/s13540-023-00192-x

Chapter 10
Attractors of Scalar Caputo FDEs

Abstract This chapter focuses on the qualitative properties of solutions to scalar dissipative Caputo fractional differential equations. The existence of attractors is demonstrated and the heteroclinic trajectories are analyzed in detail. A bifurcation diagram for simple scalar equations and triangular systems is presented.

Keywords Scalar dissipative Caputo fractional differential equations · Attractor · Heteroclinic trajectory · Bifurcation diagram

10.1 Scalar Caputo Fractional Differential Equations

Consider the scalar Caputo fractional differential equation

$$^{C}D_{0+}^{\alpha}x(t) = g(x(t)), \qquad t > 0, \tag{10.1}$$

where $g : \mathbb{R} \to \mathbb{R}$ is a continuously differentiable function and satisfies

Assumption 10.1 (Dissipative condition) There exist $a, b > 0$ such that

$$g(x)x \leq a - bx^2 \quad \text{for all } x \in \mathbb{R}.$$

Assumption 10.2 (Non-degenerate condition) $g'(x) \neq 0$ for all $x \in \mathcal{N}(g) := \{x \in \mathbb{R} : g(x) = 0\}$.

Remark 10.1 By Assumption 10.1, $\mathcal{N}(g) \subseteq \left[-\sqrt{a/b}, \sqrt{a/b}\right]$. By Assumption 10.2, $\mathcal{N}(g)$ has no accumulation point and therefore $\mathcal{N}(g)$ has a finite elements, the number of which is odd. Let $\mathcal{N}(g) = \{x_1, \ldots, x_{2k+1}\}$. Then, for all $i = 0, 1, \ldots, k$,

$$g(x) > 0 \quad \text{for all } x \in (x_{2i}, x_{2i+1}), \tag{10.2}$$

T. S. Doan et al., *Attractors of Caputo Fractional Differential Equations*, SpringerBriefs in Mathematics, https://doi.org/10.1007/978-3-032-05511-8_10

and

$$g(x) < 0 \quad \text{for all } x \in (x_{2i+1}, x_{2i+2}), \tag{10.3}$$

with the convention that $x_0 := -\infty$ and $x_{2k+2} = \infty$.

The main result of this section is the following theorem about attractors for a scalar dissipative Caputo fractional differential equation (10.1). Essentially, it says that the Caputo attractor, i.e., the compact maximal attractor given in Sect. 9.3, is the same as the attractor (as a set) of the corresponding ODE.

Theorem 10.1 *Consider the scalar Caputo fractional differential equation* (10.1). *Suppose that the Assumptions 10.1 and 10.2 hold. Then,*

(i) *The global attractor attracting all solutions starting from bounded sets is*

$$\mathcal{A} = [\min \mathcal{N}(g), \max \mathcal{N}(g)].$$

(ii) *Each solution of* (10.1) *converges to an element of* $\mathcal{N}(g)$ *and the rate of convergence is* $t^{-\alpha}$.

(iii) *Each pair of successive values of* $\mathcal{N}(g)$ *has a heteroclinic solution of* (10.1).

To prove the above theorem, several preparatory results are needed.

10.1.1   Convergence Properties

The following lemma indicates that the set $A = [x_1, x_{2k+1}]$ attracts all solutions of (10.1). Note that this set includes all steady states $x_1, \ldots, x_{2k+1}$.

Recall that the Hausdorff semi-distance between two subsets A and B of $\mathbb{R}$ is defined by

$$\text{dist}_{\mathbb{R}}(A, B) := \sup_{a \in A} \text{dist}_{\mathbb{R}}(a, B), \quad \text{dist}_{\mathbb{R}}(a, B) := \inf_{b \in B} |a - b|.$$

Lemma 10.1 *Let B be a bounded set of $\mathbb{R}$ and let $x(t, B) := \cup \{x(t, \eta) : \eta \in B\}$ for any $t \geq 0$, where $x(\cdot, \eta)$ is the solution of* (10.1) *with the initial condition $x(0) = \eta$. Then, there exists $\lambda > 0$ such that*

$$\text{dist}_{\mathbb{R}}(x(t, B), [x_1, x_{2k+1}]) \leq E_\alpha(-\lambda t^\alpha)\text{dist}_{\mathbb{R}}(B, [x_1, x_{2k+1}]) \quad \textit{for all } t \geq 0.$$

Consequently,

$$\lim_{t \to \infty} \text{dist}_{\mathbb{R}}(x(t, B), [x_1, x_{2k+1}]) = 0.$$

Proof Due to the non-intersection of two trajectories of (10.1),

$$x(t, \eta) \in [x(t, \inf B), x(t, \sup B)] \quad \text{for all } \eta \in B.$$

Then, to conclude the proof it suffices to show that for all $\eta \in \mathbb{R}$ there exists $\gamma > 0$ such that

$$\text{dist}_{\mathbb{R}}(x(t, \eta), [x_1, x_{2k+1}]) \leq E_\alpha(-\gamma t^\alpha)\text{dist}_{\mathbb{R}}((\eta, [x_1, x_{2k+1}]). \tag{10.4}$$

Using the non-intersection of two trajectories of (10.1), the fact that x_1, x_{2k+1} are steady state solutions gives

$$\text{dist}_{\mathbb{R}}(x(t, \eta), [x_1, x_{2k+1}]) = \text{dist}_{\mathbb{R}}((\eta, [x_1, x_{2k+1}]) = 0$$

for all $\eta \in [x_1, x_{2k+1}]$, which implies that (10.4) holds. Then, it is enough to deal with the case that $\eta < x_1$ and use analogous arguments for the case $\eta > x_{2k+1}$. Choose and fix $\eta < x_1$. To conclude the proof it needs to be shown that

$$|x(t, \eta) - x_1| \leq E_\alpha(-\gamma t^\alpha)|x_1 - \eta| \quad \text{for } t \geq 0. \tag{10.5}$$

for some $\gamma > 0$. The proof is divided into two steps:

Step 1: Consider a new fractional differential equation

$$^{C}\!D^\alpha_{0+}y(t) = f(y(t)), \tag{10.6}$$

where $f : \mathbb{R} \to \mathbb{R}$ is defined as

$$f(x) := g(x + x_1) \quad \text{for all } x \in \mathbb{R}. \tag{10.7}$$

Then, $x(t, \eta) = x_1 + y(t, \eta - x_1)$, where $y(\cdot, \zeta)$ denotes the solution of (10.6) satisfying $y(0) = \zeta$. To see this recall that the integral form of (10.1) is

$$x(t, \eta) = \eta + \frac{1}{\Gamma(\alpha)} \int_0^t (t - s)^{\alpha-1} g(x(s, \eta)) \, ds.$$

Thus,

$$x(t, \eta) - x_1 = \eta - x_1 + \frac{1}{\Gamma(\alpha)} \int_0^t (t - s)^{\alpha-1} g(x(s, \eta) - x_1 + x_1) \, ds$$

$$= \eta - x_1 + \frac{1}{\Gamma(\alpha)} \int_0^t (t - s)^{\alpha-1} f(x(s, \eta) - x_1) \, ds,$$

which implies that $y(t, x_1 - \eta) = x(t, \eta) - x_1$ for all $t \geq 0$. So, to prove (10.5) it suffices to show that

$$|y(t, \zeta)| \leq E_\alpha(-\gamma t^\alpha)|\zeta| \qquad \text{for} \quad \zeta := \eta - x_1 < 0. \tag{10.8}$$

Step 2: To clarify (10.8), it will first be shown that that there exists $\gamma > 0$ such that

$$f(x) \geq \gamma |x| \qquad \text{for all } x \in [2\zeta, 0]. \tag{10.9}$$

Indeed, by Assumption 10.1, $g(x) > 0$ for $x < x_1$ and therefore, by Assumption 10.2 $g'(x_1) < 0$. Equivalently, by (10.7), $f(x) > 0$ for all $x < 0$, $f(0) = 0$ and $f'(0) < 0$. Hence, by the Mean Value Theorem, there exists an $\varepsilon > 0$ such that

$$|f(x)| \geq \frac{1}{2} \min_{\xi \in [-\varepsilon, \varepsilon]} |f'(\xi)||x| \qquad \text{for all } x \in [-\epsilon, \epsilon].$$

Thus, if $x \geq -\epsilon$, then (10.9) holds for $\gamma := \frac{1}{2} \min_{\xi \in [-\varepsilon, \varepsilon]} |f'(\xi)|$. In the other case, i.e., $x < -\epsilon$, let

$$\gamma := \min \left\{ \frac{1}{2} \min_{\xi \in [-\varepsilon, \varepsilon]} |f'(\xi)|, \ \min_{w \in [2\zeta, -\epsilon]} \frac{f(w)}{|w|} \right\}.$$

Then $\gamma > 0$ by the strict positivity of f on $[\zeta, -\epsilon]$. Obviously (10.9) also holds for this choice of γ. So, in both cases there exists a $\gamma > 0$ satisfying (10.9).

Now rewrite (10.2) in the following form

$$^C D_{0+}^\alpha y(t) = -\gamma y(t) + h(y(t)),$$

where $h : \mathbb{R} \to \mathbb{R}$ is defined by

$$h(y) := f(y) + \gamma y.$$

Then (10.9) gives

$$h(y) \geq 0 \qquad \text{for all } y \in [2\zeta, 0]. \tag{10.10}$$

The variation of constants formula (see, e.g., [1, Lemma 3.1]) yields the following representation of the solution $y(t, \zeta)$ as

$$y(t, \zeta) = E_\alpha(-\gamma t^\alpha)\zeta + \frac{1}{\Gamma(\alpha)} \int_0^t (t - s)^{\alpha-1} E_{\alpha,\alpha}(-\gamma(t - s)^\alpha) h(y(s, \zeta)) \, ds. \tag{10.11}$$

By the non-intersection of two solutions of (10.2), $y(t, \zeta) < 0$ for all $t \geq 0$. Together with (10.10) and (10.11) this gives $y(t, \zeta) \in [\zeta, 0)$ for all $t \geq 0$ and

$$h(y(s, \zeta)) \geq 0 \qquad \text{for all } s \geq 0.$$

Consequently, (10.11) and the positivity of the function $E_{\alpha,\alpha}$ lead to

$$y(t, \zeta) \in [E_\alpha(-\gamma t^\alpha)\zeta, 0] \qquad \text{for all } t \geq 0,$$

which establishes (10.8). Moreover, $\lim_{t\to\infty} y(t, \zeta) = 0$ since $\lim_{t\to\infty} E_\alpha(-\gamma t^\alpha) = 0$. This completes the proof of Lemma 10.1. $\qquad\qquad\square$.

The following result establishes the asymptotic behaviour of solutions starting inside the attractor. The idea of the proof is quite similar to Lemma 10.1, so only the main points of the proof are indicated here.

Lemma 10.2 *The following statements hold:*

(i) For $i = 0, \ldots, k$ and $\eta \in (x_{2i}, x_{2i+1})$ there exists $\gamma > 0$ such that

$$|x(t, \eta) - x_{2i+1}| \leq E_\alpha(-\gamma t^\alpha)|\eta - x_{2i+1}|.$$

Consequently, $\lim_{t\to\infty} x(t, \eta) = x_{2i+1}$.

(ii) For $i = 0, \ldots, k - 1$ and $\eta \in (x_{2i+1}, x_{2i+2})$ there exists $\gamma > 0$ such that

$$|x(t, \eta) - x_{2i+1}| \leq E_\alpha(-\gamma t^\alpha)|\eta - x_{2i+1}|.$$

Consequently, $\lim_{t\to\infty} x(t, \eta) = x_{2i+1}$.

Proof Only (i) will be proved here. Analogous arguments can be used to prove (ii). Let $i \in \{0, 1, \ldots, k\}$ be arbitrary, but fixed. From (10.2) and (10.3),

$$g(x) > 0 \quad \text{for all } x \in (x_{2i}, x_{2i+1}) \quad \text{and } g'(x_{2i+1}) < 0. \tag{10.12}$$

Now, choose and fix an arbitrary $\eta \in (x_{2i}, x_{2i+1})$. Consider the new fractional differential equation

$$^{C}D^\alpha_{0+} y(t) = f(y(t)), \tag{10.13}$$

where $f : \mathbb{R} \to \mathbb{R}$ is defined as

$$f(y) := g(y + x_{2i+1}) \qquad \text{for all } x \in \mathbb{R}.$$

Let $y(t, \zeta)$ denote the solution of (10.13) satisfying $y(0, \zeta) = \zeta$. Then $x(t, \eta) = y(t, \eta - x_{2i+1}) + x_{2i+1}$ for all $t \geq 0$. It is sufficient to show that for all $\zeta \in (x_{2i} - x_{2i+1}, 0)$

$$|y(t, \zeta)| \leq E_\alpha(-\gamma t^\alpha)|\zeta| \qquad \text{for some } \gamma > 0. \tag{10.14}$$

The property (10.12) translates to the function f as

$$f(y) > 0 \quad \text{for all } y \in (x_{2i} - x_{2i+1}, 0) \quad \text{and } f'(0) < 0,$$

from which it follows that there exists $\gamma > 0$ (depending on $\zeta \in (x_{2i} - x_{2i+1}, 0)$) such that $f(y) \geq \gamma |y|$ for all $y \in [\zeta - \varepsilon, 0]$, where $\varepsilon > 0$ small enough satisfying $\zeta - \varepsilon + x_{2i+1} > x_{2i}$. Thus, by the variation of constants formula,

$$y(t, \zeta) \geq E_\alpha(-\gamma t^\alpha)\zeta \qquad \text{for all } t \geq 0,$$

which proves (10.14) and finishes the proof of Lemma 10.2. $\qquad\qquad\square$

10.1.2   Heteroclinic Trajectories

Trajectories joining distinct steady state solutions are called heteroclinic. This requires the corresponding solutions to be defined for all time. Solutions of Caputo FDEs are, however, given only forward in time, so and need to be extended to the negative time axis.

Roughly speaking, the solution $x(\cdot, \eta)$ of a scalar Caputo FDE can be extended to the negative time axis as follows: for any $\tau \leq 0$ then $x(\tau, \eta)$ is the unique value $\zeta \in \mathbb{R}$ satisfying that $x(-\tau, \zeta) = \eta$, hence $x(-\tau, x(\tau, \eta)) = \eta$. In terms of the two-parameter flow in Chap. 8 it is written $\Phi_{0,-\tau}^{-1}(\eta) = \zeta$.

In particular, this involve solving the end time problem

$$^{C}D_{0+}^{\alpha}x(t) = g(x(t)), \quad x(-\tau) = \eta$$

on the interval $0 \leq t \leq -\tau$ using Theorem 3.10. The uniform boundedness condition (3.25) there holds by restricting the vector field g to the invariant closed and bounded set between successive steady state solution. The uniqueness of ζ given τ and η then follows by the non-intersection results in Chaps. 7 and 8.

Theorem 10.2 *The following statements hold:*

(i) *For $i = 0, \ldots, k - 1$ and $\eta \in (x_{2i}, x_{2i+1})$ the solution $x(t, \eta)$ is a heteroclinic trajectory joining the steady states x_{2i} and x_{2i+1}, i.e., with*

$$\lim_{t \to -\infty} x(t, \eta) = x_{2i} \text{ and } \lim_{t \to \infty} x(t, \eta) = x_{2i+1}.$$

(ii) *For $i = 0, \ldots, k - 1$ and $\eta \in (x_{2i+1}, x_{2i+2})$ the solution $x(t, \eta)$ is a heteroclinic trajectory joining the steady states x_{2i+1} and x_{2i+2}. i.e., with*

$$\lim_{t \to -\infty} x(t, \eta) = x_{2i+2} \text{ and } \lim_{t \to \infty} x(t, \eta) = x_{2i+1}.$$

Proof Only a proof for the part (i) is given here, analogous arguments can be used for the proof of part (ii). In fact, by Lemma 10.2 it is only required to prove that

$$\lim_{t \to -\infty} x(t, \eta) = x_{2i} \quad \text{for all } \eta \in (x_{2i}, x_{2i+1}). \tag{10.15}$$

Similarly to the proofs of Lemmas 10.1 and 10.2 (i), the system can be transformed so that $x_{2i+1} = 0$. Let $x_{2i+1} = 0$. Choose and fix $\eta \in (x_{2i}, 0)$. The remaining proof into several steps.

Step 1: It will be shown that there exists $\gamma > 0$ such that

$$g(\zeta) \geq \gamma(\zeta - x_{2i})|\zeta| \quad \text{for all } \zeta \in [x_{2i}, 0]. \tag{10.16}$$

To prove this, since $g'(x_{2i}) > 0 > g'(0)$, it follows that there exists $\epsilon \in (0, -\frac{x_{2i}}{3})$ such that

$$g(\zeta) \geq \min_{\xi \in [x_{2i}-\varepsilon, x_{2i}+\varepsilon]} \frac{g'(\xi)}{2}(\zeta - x_{2i}) \quad \text{for all } \zeta \in (x_{2i}, x_{2i} + \epsilon).$$

and

$$g(\zeta) \geq \min_{\xi \in [-\varepsilon, \varepsilon]} \frac{|g'(\xi)|}{2}|\zeta| \quad \text{for all } \zeta \in (-\epsilon, 0).$$

Then, (10.16) holds for

$$\gamma := \min \left\{ \frac{\min_{\xi \in [x_{2i}-\varepsilon, x_{2i}+\varepsilon]} g'(\xi)}{2|x_{2i}|}, \frac{\min_{\xi \in [-\varepsilon, \varepsilon]} |g'(\xi)|}{2|\epsilon + x_{2i}|}, \min_{\zeta \in [x_{2i}+\epsilon, -\epsilon]} \frac{g(\zeta)}{(\zeta - x_{2i})|\zeta|} \right\}.$$

The positivity of γ then follows from the fact that $g(\zeta) > 0$ for all $\zeta \in [x_{2i} + \epsilon, -\epsilon]$.

Step 2: For any $\zeta \in x_{2i}, 0)$, we show that

$$x(t, \zeta) \geq E_\alpha(-\gamma(\zeta - x_{2i})t^\alpha)\zeta \quad \text{for all } t \geq 0. \tag{10.17}$$

To show this inequality, choose and fix $\zeta \in (x_{2i}, 0)$ and let $\widehat{\gamma} := \gamma(\zeta - x_{2i})$. The Caputo fractional differential equation (10.1) can be rewritten as

$${}^{C}D_{0+}^{\alpha} x(t) = -\widehat{\gamma}x(t) + (g(x(t)) + \widehat{\gamma}x(t)).$$

Notice that due to the non-intersection of two solutions of the equation, we see that $x(t, \zeta) < 0$ for all $t \geq 0$. By the variation of constants formula (see Lemma 3.5)

$$x(t, \zeta) = E_\alpha(-\widehat{\gamma}t^\alpha)\zeta$$
$$+ \frac{1}{\Gamma(\alpha)} \int_0^t (t - s)^{\alpha-1} E_{\alpha,\alpha}(-\widehat{\gamma}(t - s)^\alpha)(g(x(s, \zeta)) + \widehat{\gamma}x(s, \zeta)) \, ds.$$

This together with (10.16) gives $x(t, \zeta) \in (x_{2i}, 0)$ for all $t \geq 0$. Applying (10.16) again shows that

$$g(x(s, \zeta)) + \widehat{\gamma} x(s, \zeta) \geq \gamma(\zeta - x_{2i})|x(s, \zeta)| + \widehat{\gamma} x(s, \zeta) \geq 0, \; \forall s \geq 0,$$

and thus $x(t, \zeta) \geq E_\alpha(-\widehat{\gamma} t^\alpha)\zeta$.

Step 3: Let $\eta \in (x_{2i}, 0)$ be arbitrary. Then,

$$x(-t, x(t, \eta)) = \eta \qquad \text{for all } t < 0,$$

which together with (10.17) implies that

$$\begin{aligned}
\eta &\geq E_\alpha(-\gamma(x(t, \eta) - x_{2i})(-t)^\alpha)x(t, \eta) \\
&\geq E_\alpha(-\gamma(x(t, \eta) - x_{2i})(-t)^\alpha)x_{2i} \quad \text{for all } t < 0.
\end{aligned} \tag{10.18}$$

It is worth noting that $E_\alpha(\cdot)$ is a montononically increasing function and

$$\lim_{t \to -\infty} E_\alpha(-\rho(-t)^\alpha) = 0 \quad \text{for all } \rho > 0. \tag{10.19}$$

Suppose the contrary that $\overline{\lim}_{t \to -\infty} x(t, \eta) - x_{21} = a > 0$. From (10.19), for any $\varepsilon > 0$ small enough, we can find $T > 0$ such that $E_\alpha(-\gamma(x(t, \eta) - x_{2i})(-t)^\alpha) < \varepsilon$ for all $t < -T$. This means that $\lim_{t \to -\infty} E_\alpha(-\gamma(x(t, \eta) - x_{2i})(-t)^\alpha) = 0$, which combining with (10.18) leads to $\eta \geq 0$, a contradiction. Therefore $\lim_{t \to -\infty} x(t, \eta) - x_{2i} = 0$, that is, $\lim_{t \to -\infty} x(t, \eta) = x_{2i}$. The proof is complete. $\qquad \square$

10.1.3   Proof of Theorem 10.1

The above Lemmas and Theorem can be combined to complete the proof of Theorem 10.1.

Proof The proofs of (i) and (iii) are given in Lemma 10.1 and Theorem 10.2, respectively. The proof of the first statement in (ii) that each solution of (10.1) converges to an element of $\mathcal{N}(g)$ is given in Lemma 10.2.

It remains to show the rate of convergence. In fact, using Lemma 10.2, for any η there exists $\gamma > 0$ such that

$$\text{dist}_{\mathbb{R}}(x(t, \eta), \mathcal{N}(g)) \leq E_\alpha(-\gamma t^\alpha)\text{dist}_{\mathbb{R}}(\eta, \mathcal{N}(g)) \qquad \text{for all } t \geq 0. \tag{10.20}$$

On the other hand, by [1, Theorem 4.1] there exists $L > 0$ such that

$$\text{dist}_{\mathbb{R}}(x(t, \eta), \mathcal{N}(g)) \geq E_\alpha(-L t^\alpha)\text{dist}_{\mathbb{R}}(\eta, \mathcal{N}(g)) \qquad \text{for all } t \geq 0. \tag{10.21}$$

Furthermore, $\lim_{t\to\infty} t^\alpha E_\alpha(-\lambda t^\alpha)$ is finite for any $\lambda > 0$. Then, using (10.20) and (10.21), the rate of convergence of any solution of (10.1) to the steady states of (10.1) is $t^{-\alpha}$.

This completes the proof of Theorem 10.1. $\square$

10.1.4 Comparison with the Corresponding Scalar ODEs

Consider a scalar ordinary differential equation

$$\dot{x}(t) = g(x(t)). \tag{10.22}$$

Let a, b be two successive zeros of g, i.e., $g(a) = g(b) = 0$ and $g(x) \neq 0$ for $x \in (a, b)$. Then, by continuity of g either $g(x) > 0$ for all $x \in (a, b)$ or either $g(x) < 0$ for all $x \in (a, b)$. Therefore, any solution starting from a value in (a, b) will be either strictly monotonically increasing or strictly monotonically decreasing. Consequently, any solution of (10.22) will converges to one of two steady states a, b.

The above monotonicity argument for ODEs cannot extend to Caputo FDEs with the same vector field g. The main reason is the appearance of the singular kernel in the integral form

$$x(t, \eta) = \eta + \frac{1}{\Gamma(\alpha)} \int_0^t (t - s)^{\alpha-1} g(x(s, \eta)) \, ds.$$

Example 10.1 Two specific scalar Caputo FDE, i.e., with a vector field $g : \mathbb{R} \to \mathbb{R}$, namely will be investigated:.

$$g(x) = -x, \quad g(x) = x - x^3.$$

These satisfy a dissipativity condition and have steady state solutions 0 and $0, \pm 1$, respectively. The ODEs

$$\frac{d}{dt} x(t) = g(x(t))$$

with these vector fields have global attractors $\mathcal{A} = \{0\}$ and $\mathcal{A} = [-1, 1]$, respectively. Then, the corresponding Caputo FDEs

$$^C D_{0+}^\alpha x(t) = g(x(t))$$

have the same steady state solutions and attractors, see Fig. 10.1. A major difference is that attraction or repulsion of the steady state solutions is not at an exponential rate in the Caputo case.

Also, in the second example, the heteroclinic trajectories joining the steady state solutions have the same geometric image in $\mathbb{R}$, but have different functional representations in the ODE and Caputo systems.

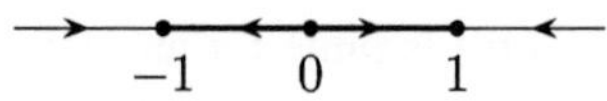

Fig. 10.1 Attractor $\mathcal{A} = [-1, 1]$ for $g(x) = x - x^3$

10.2   Bifurcations for Scalar Caputo FDEs

Consider a family of scalar Caputo FDEs (10.1)

$$^{C}D_{0+}^{\alpha} x(t) = g(\gamma, x),\tag{10.23}$$

where γ is a parameter. Let $x_\gamma(t, \eta)$ denote the solution of (10.23) satisfying $x(0) = \eta$. It follows from Theorem 10.1 that (10.23) has the same bifurcations as an ODE with the same vector field. This means, in particular, that the simpler steady states and sign of vector fields for the ODE can be used to determine the bifurcations of the corresponding Caputo FDE.

In what follows, the saddle-node and pitchfork bifurcations are investigated for scalar Caputo FDEs. Readers are referred to [2, Sect. 2.1] for a corresponding bifurcation analysis of ODE.

Example 10.2   (Saddle-node bifurcation) Consider the following family of scalar Caputo FDEs

$$^{C}D_{0+}^{\alpha} x(t) = \gamma - x(t)^2.\tag{10.24}$$

Then, the following statements hold:

(i) For $\gamma < 0$, then all solutions of (10.24) tends to $-\infty$.

(ii) For $\gamma \geq 0$, then (10.24) has two steady states $x = -\sqrt{\gamma}$ and $x = \sqrt{\gamma}$ and

$$\lim_{t \to \infty} x_\gamma(t, \eta) = \begin{cases} -\infty, & \text{if } \eta < -\sqrt{\gamma}; \\ \sqrt{\gamma}, & \text{if } \eta > -\sqrt{\gamma}. \end{cases}$$

An analytical proof for (i) comes from the fact that

$$x_\gamma(t, \eta) = \eta + \frac{1}{\Gamma(\alpha)} \int_0^t (\gamma - x_\gamma(s, \eta)^2)\, ds \leq \eta + \frac{\gamma t}{\Gamma(\alpha)}.$$

For the case $\gamma \geq 0$, an analogous argument as in (i) implies that $\lim_{t \to \infty} x_\gamma(t, \eta) = -\infty$ for $\eta < -\sqrt{\gamma}$.

Moreover, using Theorem 10.1 for the restriction of (10.24) on $(-\sqrt{\gamma}, \infty)$ leads to $\lim_{t \to \infty} x_\gamma(t, \eta) = \sqrt{\gamma}$ if $\eta > -\sqrt{\gamma}$.

Example 10.3 (Pitchfork bifurcation) Consider the following family of scalar Caputo FDEs

$$^{C}D_{0+}^{\alpha}x(t) = \gamma x(t) - x(t)^{3}. \tag{10.25}$$

Then Theorem 10.1 leads to the following description of the bifurcation of the asymptotical behaviour of solutions of (10.25) on the parameter γ:

(i) For $\gamma < 0$, then (10.24) has a steady state $x = 0$ attracting all solutions of (10.24).

(ii) For $\gamma \geq 0$, then (10.24) has three steady states $x = -\sqrt{\gamma}, x = 0, x = \sqrt{\gamma}$ and

$$\lim_{t \to \infty} x_{\gamma}(t, \eta) = \begin{cases} -\sqrt{\gamma}, & \text{if } \eta < 0, \\ \sqrt{\gamma}, & \text{if } \eta > 0. \end{cases}$$

10.3 Caputo FDEs with Triangular Vector Fields

The result in previous section are generalised to a special class of Caputo fractional differential equations with triangular vector fields of the following form

$$^{C}D_{0+}^{\alpha}x(t) = g(x(t)) = (g_{1}(x(t)), \ldots, g_{d}(x(t)))^{\mathrm{T}}, \tag{10.26}$$

where for $i = 1, \ldots, d$ it is assumed that the function $g_{i} : \mathbb{R}^{d} \to \mathbb{R}$ is of the following form

$$g_{i}(x) = h_{i}(x_{1}, \ldots, x_{i-1}) f_{i}(x_{i}) \qquad \text{for } i = 1, \ldots, d.$$

The function g is also assumed to be continuously differentiable and to satisfy the following hypothesises.

Assumption 10.3 (Dissipative condition) There exist $a, b > 0$ such that

$$\langle x, g(x) \rangle \leq a - b\|x\|^{2} \quad \text{for all } x \in \mathbb{R}^{d}.$$

Assumption 10.4 (Non-degenerate condition) $g_{i}'(u) \neq 0$ for all $u \in \mathcal{N}(g_{i}) := \{u \in \mathbb{R} : g_{i}(u) = 0\}$.

Since the structure of vector field in (10.26) is of product form it follows with the assume (H1) that for all $i = 1, \ldots, d$ the function $h_{i} : \mathbb{R}^{i-1} \to \mathbb{R}$ does not vanishing. Thus, a change of sign of g_{i} depends only on a change of sign of the function f_{i}. Applying Theorem 10.1 to each component thus leads to the following result on attractors for Caputo fractional differential equations of triangular vector fields.

Theorem 10.3 *Consider system (10.26). Suppose that the Assumptions 10.3 and 10.4 hold. Then, the following statements hold:*

(i) The global attractor attracting all solutions starting from bounded sets is

$$\mathcal{A} = [\min \mathcal{N}(g_{11}), \max \mathcal{N}(g_{11})] \times \cdots \times [\min \mathcal{N}(g_{dd}), \max \mathcal{N}(g_{dd})].$$

(ii) Each solution of (10.1) converges to an element of the following set

$$\mathcal{N}(g) := \mathcal{N}(g_{11}) \times \cdots \times \mathcal{N}(g_{dd})$$

and the rate of convergence is $t^{-\alpha}$.

Example 10.4 Consider the two-dimensional triangular Caputo FDEs with the following vector field

$$^{C}D_{0+}^{\alpha}x(t) = x(t)(1 - x(t)), \quad ^{C}D_{0+}^{\alpha}y(t) = y(t)(1 - y(t)^2)(1 + x(t)^2).$$

The attracting set for the above equation is $\mathcal{A} = [-1, 1] \times [-1, 1]$. The asymptotically behaviour of solutions is depicted in the Fig. 10.2.

Remark 10.2 It would be interesting to know whether or not Theorem 10.3 remains true when the vector field is of a more general form of triangular vector fields, e.g.,

$$^{C}D_{0+}^{\alpha}x(t) = f(x(t)), \quad ^{C}D_{0+}^{\alpha}y(t) = g(x(t), y(t)).$$

The non-intersection result of two solutions considered in Chap. 7 is still true for this equation. The sign of the vector field $g(x(t), y(t))$, however, depends on both $x(t)$ and $y(t)$, so the approach used above does not apply here.

Fig. 10.2 Attractor $\mathcal{A} = [-1, 1] \times [-1, 1]$ with steady state and heteroclinic trajectories for the vector field $g(x) = x(1 - x^2)$, $f(x, y) = y(1 - y^2)(1 + x^2)$

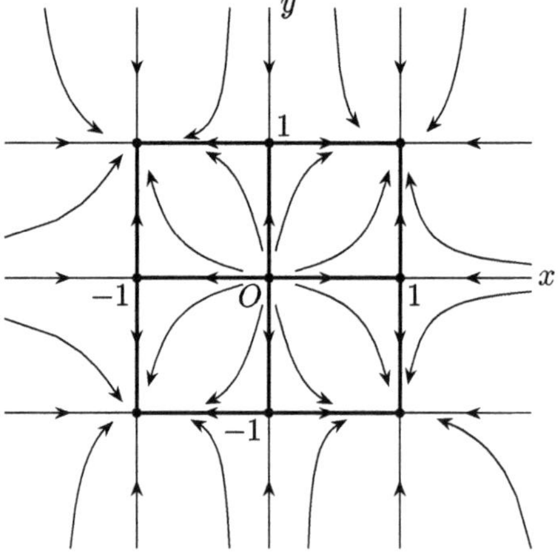

10.4 Endnotes

This chapter is based on Doan and Kloeden [3]. See Hale [4] for more information on autonomous semi-dynamical systems and their attractors, and Kloeden and Yang [5] for the non autonomous case. Bifurcations for autonomous ODEs are discussed, e.g., in Hale and Koçak [2].

References

1. Cong, N.D., Tuan, H.T.: Generation of nonlocal dynamical systems by fractional differential equations. J. Integral Equ. Appl. **29**, 585–608 (2017). https://doi.org/10.1216/JIE-2017-29-4-585
2. Hale, J.K., Koçak, H.: Dynamics and Bifurcations. Springer-Verlag, New York (1991)
3. Doan, T.S., Kloeden, P.E.: Attractors of Caputo fractional differential equations with triangular vector field. Fract. Calc. Anal. Appl. **25**, 720–734 (2022). https://doi.org/10.1007/s13540-022-00030-6
4. Hale, J.K.: Asymptotic behavior of dissipative system. American Mathematical Society, Providence, Rhode Island (1988)
5. Kloeden, P.E., Yang, M.: Introduction to nonautonomous dynamical systems and their attractors. World Scientific Publishing Co., Inc, Singapore (2021)

Chapter 11
Attractors of Caputo Semi-Groups

Abstract The existence of attractors for operator semi-groups induced by a dissipative Caputo fractional differential equation is established on a space of continuous functions, endowed with the topology of uniform convergence on compact subsets.

Keywords Dissipative Caputo fractional differential equations · Attractors · Caputo semi–group of operators

11.1 Introduction

Consider an autonomous Caputo FDE of order $\alpha \in (0, 1)$ in $\mathbb{R}^d$ of the form

$$^{C}D_{0+}^{\alpha}x(t) = g(x(t)) \tag{11.1}$$

where $g : \mathbb{R}^d \to \mathbb{R}^d$ is locally Lipschitz continuous and satisfies a growth bound. The Caputo FDE (11.1) with the initial condition $x(0) = x_0$ is essentially the integral equation

$$x(t) = x_0 + \frac{1}{\Gamma(\alpha)} \int_0^t (t - s)^{\alpha-1} g(x(s)) ds. \tag{11.2}$$

Let $\mathfrak{C}$ be the Banach space of continuous functions $f : [0, \infty) \to \mathbb{R}^d$ with the topology uniform convergence on compact subsets, which has the metric

$$\rho(f, h) := \sum_{n=1}^{\infty} \frac{1}{2^n} \rho_n(f, h), \quad \text{where } \rho_n(f, h) := \frac{\sup_{t \in [0,n]} \|f(t) - h(t)\|}{1 + \sup_{t \in [0,n]} \|f(t) - h(t)\|}. \tag{11.3}$$

Given $f \in \mathfrak{C}$, define the operators $T_t : \mathfrak{C} \to \mathfrak{C}$, $t \geq 0$, by

$$(T_t f)(\theta) = f(t + \theta) + \frac{1}{\Gamma(\alpha)} \int_0^t (t + \theta - s)^{\alpha-1} g(x_f(s)) \, ds, \qquad \theta \geq 0, \tag{11.4}$$

T. S. Doan et al., *Attractors of Caputo Fractional Differential Equations*, SpringerBriefs in Mathematics, https://doi.org/10.1007/978-3-032-05511-8_11

where x_f is a solution of the singular Volterra integral equation for this f, i.e.,

$$x_f(t) = f(t) + \frac{1}{\Gamma(\alpha)} \int_0^t (t - s)^{\alpha-1} g(x_f(s))\, ds. \tag{11.5}$$

It was seen in Chap. 8 that the operators T_t, $t \geq 0$, form a semi-group on the space $\mathfrak{C}$. This semi-group represents the Caputo FDE (11.1) as an autonomous semi-dynamical system on the space $\mathfrak{C}$ when $f(t) \equiv id_{x_0}$ for $x_0 \in \mathbb{R}^d$.

11.2 Dissipative Vector Fields

Suppose that the vector field g of the Caputo FDE (11.1) satisfies a dissipativity condition

$$\langle x, g(x) \rangle \leq a - b\|x\|^2, \tag{11.6}$$

where $a, b > 0$. It was shown in Chap. 9 using the Tuan and Trinh inequality (2.5) that the solutions of the Caputo FDE (11.1) satisfy the inequality

$$\|x(t, x_0)\|^2 \leq \|x_0\|^2 E_\alpha(-2bt^\alpha) + \frac{a}{b}\left(1 - E_\alpha(-2bt^\alpha)\right). \tag{11.7}$$

It follows from this that $\|x(t, x_0)\| \leq R$ for all $t \geq 0$ when $\|x_0\| \leq R$ and $R^2 \geq 1 + \frac{a}{b}$ $=: R_*^2$ and that the set

$$\mathcal{B}^* := \left\{ x \in \mathbb{R}^d \ : \ \|x\|^2 \leq R_*^2 \right\}$$

is a positively invariant absorbing set for the solutions of the Caputo FDE (11.1).

Notice that for $R^2 < 1 + \frac{a}{b}$, from (11.7), it is obvious to see that $\|x(t, x_0)\|^2 <$ $(1 + \frac{a}{b})E_\alpha(-2bt^\alpha) + \frac{a}{b}(1 - E_\alpha(-2bt^\alpha)) \leq 1 + \frac{a}{b}$ for all $t \geq 0$. Therefore, there exists $T_R \geq 0$ such that $\|x(t, x_0)\| \in \mathcal{B}^*$, i.e., $\|x(t, x_0)\| \leq R^*$ for all $t \geq T_R$.

Since this set $\mathcal{B}^*$ is compact in $\mathbb{R}^d$, the corresponding omega limit set $\Omega_{\mathcal{B}^*}$ exists and is a nonempty compact subset of $\mathcal{B}^*$. Moreover, it attracts all of the future dynamics of the Caputo FDE (11.1) and contains all of the steady state solutions.

It was seen in Chap. 9 that $\Omega_{\mathcal{B}^*}$ is the compact minimal attractor of dissipative scalar Caputo FDEs. In general, $\Omega_{\mathcal{B}^*}$ cannot be considered as the attractor of the autonomous Caputo FDE (11.1), since the corresponding semi-dynamical system is defined on the function metric space $(\mathfrak{C}, \rho)$ and not on $\mathbb{R}^d$. Nevertheless, it will seen below that $\Omega_{\mathcal{B}^*}$ represents the *observable* (in $\mathbb{R}^d$) part of an attractor $\mathfrak{A}$ in $\mathfrak{C}$ of this Caputo semi-dynamical system and, essentially, determines it.

Theorem 9.3 from theory of autonomous semi-dynamical systems [1, 2] implies the existence of a global attractor under appropriate assumptions. Unfortunately, this theorem cannot be applied to the Caputo semi-group $\{T_t\}_{t \in \mathbb{R}^+}$ since the attracting property is restricted to constant initial functions $f(t) \equiv id_{x_0}$ corresponding to initial values $x_0 \in \mathbb{R}^d$. This can be handled by restricting the sets that are attracted to an appropriate universe of sets that are attracted.

Another difficulty is how to apply the dissipativity condition (11.6) to the vector field g inside the integral equations (11.4) defining the Caputo semi-group for $\theta > 0$ to establish the existence of an absorbing set in the space $\mathfrak{C}$. It will be seen that this can also be circumvented.

Restrict now to constant initial functions $f(t) \equiv id_{x_0}$, which corresponding to initial values $x_0 \in \mathbb{R}^d$. Then the dissipativity condition (11.6) can be used in the case $\theta = 0$, which corresponds to the Caputo FDE (11.1) with the initial condition $x(0) = x_0$, using the inequality (11.7), and leads to

$$x(t, x_0) \in \mathcal{B}^* \;\Leftrightarrow\; \|x(t, x_0)\| \le R_*, \quad t \ge T_R, \; \|x_0\| \le R \tag{11.8}$$

for all $R \ge R_*$.

These bounds can then be used to estimate the integrals for the integral equations (11.4) with $\theta > 0$. Essentially, the integral equations (11.4) have a skew-product like structure with the solution of

$$x(t, x_0) = x_0 + \frac{1}{\Gamma(\alpha)} \int_0^t (t - s)^{\alpha-1} g(x(s, x_0)) \, ds \tag{11.9}$$

being fed into

$$(T_t id_{x_0})(\theta) = x_0 + \frac{1}{\Gamma(\alpha)} \int_0^t (t + \theta - s)^{\alpha-1} g(x(s, x_0)) \, ds, \qquad \theta > 0. \tag{11.10}$$

Note that for $\theta = 0$

$$(T_t id_{x_0})(0) = x(t, x_0), \quad t \ge 0.$$

For technical reasons to be revealed in the proofs below a subspace of the space $\mathfrak{C}$ will be used with a weighted norm characterising uniform convergence on bounded intervals. In particular, consider the weighted norm on $\mathfrak{C}([0, \infty), \mathbb{R}^d)$ defined

$$\|f\|_\alpha := \|f(0)\| + \sum_{N=1}^\infty \frac{1}{2^N N^\alpha} \|f\|_N,$$

where

$$\|f\|_N := \sup_{t \in [N^{-1}, N]} \|f(t)\|, \quad N = 1, 2, \cdots.$$

Let $\mathfrak{C}_\alpha$ be the subspace of $\mathfrak{C}([0, \infty), \mathbb{R}^d)$ consisting of functions f with $\|f\|_\alpha < \infty$. Then $(\mathfrak{C}_\alpha, \|\cdot\|_\alpha)$ is a Banach space and $\{T_t\}_{t \in \mathbb{R}^+}$ is a semi-group on $\mathfrak{C}_\alpha$. Finally, define

$$id_D = \big\{ id_{x_0} \in \mathfrak{C}_\alpha \, : \, x_0 \in D \big\}$$

for each bounded subset D of $\mathbb{R}^d$. The attraction for the Caputo semi-dynamical system will be with respect to the the universe of bounded subsets of $\mathfrak{C}_\alpha$ given by

$$\mathfrak{D} = \left\{ id_D \, : \, D \text{ bnd}'\text{d in } \mathbb{R}^d \right\}$$

In addition, the attractor will be a closed and bounded subset of $\mathfrak{C}_\alpha$ rather than a compact subset.

Theorem 11.1 [3, Theorem 2] *Suppose that the vector field g is locally Lipschitz and satisfies the uniform dissipativity condition* (11.6). *Then the Caputo semi-group $\{T_t\}_{t \in \mathbb{R}^+}$ on the space $\mathfrak{C}_\alpha$ corresponding to the integral equations* (11.5) *has a closed and bounded $\mathfrak{D}$-attractor $\mathfrak{A}$ in $\mathfrak{C}_\alpha$, i.e., which attracts bounded subsets of constant initial value functions $f(t) \equiv id_{x_0}$ corresponding to bounded subsets of initial values $x_0 \in \mathbb{R}^d$. In particular,*

$$\mathfrak{A} = \overline{\bigcup_{\substack{D \subset \mathbb{R}^d \\ \text{bnd}'\text{d}}} \bigcap_{t \geq s} \bigcup_{s \geq 0} T_s(id_D)}.$$

The set $\mathfrak{A}$ will be called the Caputo attractor.

The proof of the existence of the attractor in Theorem 11.1 is given in remaining sections of the chapter. It will be shown that the semi-group $\{T_t\}_{t \in \mathbb{R}^+}$ is asymptotically compact and that the closed and bounded subset $\mathfrak{B}^*$ of $\mathfrak{C}_\alpha$ defined by

$$\mathfrak{B}^* := \left\{ \chi \in \mathfrak{C}_\alpha \, : \, \|\chi\|_\alpha \leq 2R_* + \frac{B_{R_*}^g}{\alpha \Gamma(\alpha)} =: \widehat{R}_* \right\}$$

absorbs under the operators T_t bounded sets of constant initial data functions $\|id_{x_0}\|_\alpha \leq 2\|x_0\| \leq 2R$ in the time $t \geq T_R$, where $B_{R_*}^g$ is a constant defined below.

Note that the absorbing set $\mathcal{B}^*$ and omega limit set $\Omega_{\mathcal{B}^*}$ in $\mathbb{R}^d$ satisfy

$$\mathcal{B}^* = \left\{ \chi(0) \in \mathbb{R}^d \, : \, \chi \in \mathfrak{B}^* \right\}, \qquad \Omega_{\mathcal{B}^*} = \left\{ \chi(0) \in \mathbb{R}^d \, : \, \chi \in \mathfrak{A} \right\}.$$

11.3 Proof of Theorem 11.1

Let $x(t, x_0)$ be the solution of the Caputo FDE (11.1) satisfying the dissipativity condition (11.6) with the initial condition $x(0, x_0) = x_0$. This solution is absorbed in the the compact subset $\mathcal{B}^*$ of $\mathbb{R}^d$ in a finite time and satisfies the bounds (11.8). Then following bounds hold.

$$B_R := \sup_{t \geq 0, \|x_0\| \leq R} \|x(t, x_0)\| < \infty, \qquad B_R^g := \sup_{\|x\| \leq R} \|g(x)\| < \infty.$$

where the continuity of the vector field g has been used in the second bound. These are valid for $R = R_*$ provided $t \geq T_R$.

The next result restates Theorem 3.5 in the dissipative case considered here.

Lemma 11.1 *The solution of the integral equation (11.1) is Hölder continuous with exponent α. In particular,*

$$\|x(t+\theta, x_0) - x(t, x_0)\| \le \frac{2 B^g_{B_R}}{\alpha \Gamma(\alpha)} \theta^\alpha$$

for $\|x_0\| \le R$ and $t \ge 0$.

11.3.1 Growth Bounded

Lemma 11.2

$$\|(T_t id_{x_{0,n}})(\theta)\| \le \frac{B^g_{R^*}}{\alpha \Gamma(\alpha)} \theta^\alpha + R^*, \quad t \ge T_R, \ \|x_{0,n}\| \le R. \qquad (11.11)$$

Proof It follows from (11.9) that

$$x(t+\theta, x_{0,n}) = x_{0,n} + \frac{1}{\Gamma(\alpha)} \int_0^{t+\theta} (t+\theta-s)^{\alpha-1} g(x(s, x_{0,n})) \, ds.$$

Hence

$$
\begin{aligned}
x(t+\theta, x_{0,n}) &= x_{0,n} + \frac{1}{\Gamma(\alpha)} \int_0^{t+\theta} (t+\theta-s)^{\alpha-1} g(x(s, x_{0,n})) \, ds \\
&= x_{0,n} + \frac{1}{\Gamma(\alpha)} \int_0^{t} (t+\theta-s)^{\alpha-1} g(x(s, x_{0,n})) \, ds \\
&\quad + \frac{1}{\Gamma(\alpha)} \int_t^{t+\theta} (t+\theta-s)^{\alpha-1} g(x(s, x_{0,n})) \, ds \\
&= (T_t id_{x_{0,n}})(\theta) + \frac{1}{\Gamma(\alpha)} \int_t^{t+\theta} (t+\theta-s)^{\alpha-1} g(x(s, x_{0,n})) \, ds.
\end{aligned}
$$

It follows that

$$
\begin{aligned}
\|x(t+\theta, x_{0,n}) - (T_t id_{x_{0,n}})(\theta)\| &= \frac{1}{\Gamma(\alpha)} \left\| \int_t^{t+\theta} (t+\theta-s)^{\alpha-1} g(x(s, x_{0,n})) \, ds \right\| \\
&\le \frac{1}{\Gamma(\alpha)} B^g_{B_R} \left| \int_t^{t+\theta} (t+\theta-s)^{\alpha-1} ds \right| \le \frac{B^g_{B_R}}{\alpha \Gamma(\alpha)} \theta^\alpha,
\end{aligned}
$$

i.e.,

$$\|x(t+\theta, x_{0,n}) - (T_t id_{x_{0,n}})(\theta)\| \le \frac{B^g_{B_R}}{\alpha \Gamma(\alpha)} \theta^\alpha, \quad t \ge 0. \qquad (11.12)$$

Hence

$$\|(T_t id_{x_{0,n}})(\theta)\| \leq \frac{B^g_{B_R}}{\alpha\Gamma(\alpha)}\theta^\alpha + \|x(t+\theta, x_{0,n})\| \leq \frac{B^g_{B_R}}{\alpha\Gamma(\alpha)}\theta^\alpha + B_R, \quad t \geq 0. \tag{11.13}$$

Note that these estimates are uniform in $t \geq 0$.

Then, using the fact that $x(t, x_0) \in \mathcal{B}^*$, i.e., $\|x(t, x_{0,n})\| \leq R^*$ for all $t \geq T_R$ and $\|x_{0,n}\| \leq R$, gives the sharper inequality (11.11). $\qquad\square$

It follows from (11.11) that

$$\|(T_t id_{x_{0,n}})(\theta)\| \leq \frac{B^g_{R^*}}{\alpha\Gamma(\alpha)}N^\alpha + R^*, \quad t \geq T_R, \ 0 \leq \theta \leq N^\alpha. \tag{11.14}$$

11.3.2　Boundedness of θ-Derivatives

The integrand in the integral

$$\int_0^t (t+\theta-s)^{\alpha-1}\, ds, \quad \theta > 0,$$

is non-singular, so the integral is in fact a classical Riemann integral. Hence we can differentiate by the parameter θ to obtain

$$\frac{d}{d\theta}\int_0^t (t+\theta-s)^{\alpha-1}\, ds = (\alpha-1)\int_0^t (t+\theta-s)^{\alpha-2}\, ds = \theta^{\alpha-1} - (t+\theta)^{\alpha-1}.$$

Similarly,

$$\frac{d}{d\theta}\int_0^t (t+\theta-s)^{\alpha-1}g(x(s, x_{0,n}))\, ds = (\alpha-1)\int_0^t (t+\theta-s)^{\alpha-2}g(x(s, x_{0,n}))\, ds.$$

Lemma 11.3

$$\left\|\frac{d}{d\theta}(T_t id_{x_{0,n}})(\theta)\right\| \leq \frac{B^g_{B_R}}{\Gamma(\alpha)}\frac{1}{\theta^{1-\alpha}} \tag{11.15}$$

for all $t \geq 0$, $\|x_{0,n}\| \leq R$ and $\theta > 0$.

Proof

$$\left\| \frac{d}{d\theta} \int_0^t (t + \theta - s)^{\alpha-1} g(x(s, x_{0,n})) \, ds \right\| = (1 - \alpha) \left\| \int_0^t (t + \theta - s)^{\alpha-2} g(x(s, x_{0,n})) \, ds \right\|$$
$$\leq (1 - \alpha) B_{B_R}^g \left| \int_0^t (t + \theta - s)^{\alpha-2} \, ds \right|$$
$$= B_{B_R}^g \left(\theta^{\alpha-1} - (t + \theta)^{\alpha-1} \right).$$

This gives

$$\left\| \frac{d}{d\theta} (T_t i d_{x_{0,n}})(\theta) \right\| \leq \frac{B_{B_R}^g}{\Gamma(\alpha)} \left(\theta^{\alpha-1} - (t + \theta)^{\alpha-1} \right) \leq \frac{B_{B_R}^g}{\Gamma(\alpha)} \frac{1}{\theta^{1-\alpha}}$$

for all $t \geq 0$ and $\theta > 0$. $\qquad\square$

11.3.3 Applying Ascoli's Theorem

Write $\chi_n(t, \theta) := (T_t i d_{x_{0,n}})(\theta)$, so $\chi_n(0, \theta) := x_{0,n}$. By estimate (11.14),

$$\|\chi_n(t, \theta)\| \leq \frac{B_{R^*}^g}{\alpha \Gamma(\alpha)} N^\alpha + R^* \tag{11.16}$$

uniformly in $\theta \in [0, N]$ for all $N > 0$ and $t \geq T_R$.

In addition, by estimate (11.15)

$$\left\| \frac{d}{d\theta} \chi_n(t, \theta) \right\| \leq \frac{B_{B_R}^g}{\Gamma(\alpha)} \frac{1}{\theta^{1-\alpha}}, \quad t \geq 0,$$

so for any $0 < \varepsilon \ll 1$,

$$\left\| \frac{d}{d\theta} \chi_n(t, \theta) \right\| \leq \frac{B_{B_R}^g}{\Gamma(\alpha)} \frac{1}{\varepsilon^{1-\alpha}}$$

uniformly in $\theta \in [\varepsilon, \infty)$ for all $t \geq 0$. The $\chi_n(t, \cdot)$ are thus equi-Lipschitz continuous uniformly in $\theta \in [\varepsilon, \infty)$ for all $t \geq 0$.

This means the Ascoli theorem can be applied on each interval of the form $[\varepsilon, N]$, i.e., in the space $\mathfrak{C}([\varepsilon, N], \mathbb{R}^d)$ of continuous functions $f : [\varepsilon, N]) \to \mathbb{R}^d$. Thus there are (sub)sequences $t_n \to \infty$ and a function $\chi^* \in \mathfrak{C}([\varepsilon, N], \mathbb{R}^d)$ such that

$$\chi_n(\theta) := \chi_n(t_n, \theta) \to \chi^*(\theta), \quad t_n \to \infty,$$

uniformly in $\theta \in [\varepsilon, N]$ for each $N \in \mathbb{N}$.

Set $\varepsilon = N^{-1}$. Apply the above arguement on the interval $[N^{-1}, N]$ for increasing N to show that the convergent subsequence on $[N^{-1}, N]$ has a convergent subsequence on $[(N+1)^{-1}, N+1]$. Then use the diagonal subsequence to show that $\chi^*(\theta)$ is defined for all $\theta > 0$.

11.3.4   Continuity of $\chi^*(\theta)$ at $\theta = 0$

It follows from Lemma 11.1 and the dissipativity condition that

$$\left\| x(t+\theta, x_{0,n}) - x(t, x_{0,n}) \right\| \leq \frac{2B_{B_R}^g}{\alpha \Gamma(\alpha)} \theta^\alpha$$

for all $t \geq 0$ and $\theta \geq 0$. Let $t_n \geq T_R$. Then

$$\left\| x(t_n+\theta, x_{0,n}) - x(t_n, x_{0,n}) \right\| \leq \frac{2B_{R^*}^g}{\alpha \Gamma(\alpha)} \theta^\alpha$$

for all $\theta \geq 0$.

For each $\theta > 0$ there is a convergent subsequence (of the subsequence used above to obtain χ^*) such that the limits $\lim_{n\to\infty} x(t_n+\theta, x_{0,n}) = x^*(\theta)$, $\lim_{n\to\infty} x(t_n, x_{0,n}) = x^*$ exist and satisfy

$$\left\| x^*(\theta) - x^* \right\| \leq \frac{2B_{R^*}^g}{\alpha \Gamma(\alpha)} \theta^\alpha.$$

It is also clear from the estimate (11.12) that

$$\left\| x^*(\theta) - \chi^*(\theta) \right\| \leq \frac{B_{R^*}^g}{\alpha \Gamma(\alpha)} \theta^\alpha, \quad \theta \geq 0.$$

Thus

$$\left\| \chi^*(\theta) - x^* \right\| \leq \left\| \chi^*(\theta) - x^*(\theta) \right\| + \left\| x^* - x^*(\theta) \right\|$$

$$\leq \frac{3B_{R^*}^g}{\alpha \Gamma(\alpha)} \theta^\alpha \to 0 \quad \text{as } \theta \to 0.$$

Summarising,

Lemma 11.4 $\chi^* \in \mathfrak{C}_\alpha([0, \infty), \mathbb{R}^d)$.

Thus the operator $(T_t i d_{x_{0,n}})(\cdot)$ is asymptotically compact on bounded intervals, i.e., for every sequence $t_n \to \infty$ and $\|x_{0,n}\| \leq R$ and there is a subsequence $t_n \to \infty$ such that $\chi_n(t_n, \cdot) = (T_{t_n} i d_{x_{0,n}})(\cdot) \to \chi^*(\cdot) \in \mathfrak{C}([0, \infty), \mathbb{R}^d)$.

11.3.5 Estimates in the Weighted Norm

In terms of the weighted norm $\|\cdot\|_\alpha$ on $\mathfrak{C}_\alpha$, the bound (11.16) becomes

$$\|\chi_n(t,\cdot)\|_\alpha = \|\chi_n(t,0)\| + \sum_{N=1}^{\infty} \frac{1}{2^N N^\alpha} \|\chi_n(t,\theta)\|_N,$$

$$\leq R^* + \sum_{N=1}^{\infty} \frac{1}{2^N N^\alpha} \left(\frac{B_{R^*}^g}{\alpha \Gamma(\alpha)} N^\alpha + R^* \right)$$

for all $t \geq T_R$. Hence

$$\|\chi_n(t,\cdot)\|_\alpha \leq R^* \left(1 + \sum_{N=1}^{\infty} \frac{1}{2^N N^\alpha} \right) + \frac{B_{R^*}^g}{\alpha \Gamma(\alpha)} \sum_{N=1}^{\infty} \frac{1}{2^N} \leq 2R^* + \frac{B_{R^*}^g}{\alpha \Gamma(\alpha)} = \widehat{R}^*$$

for all $t \geq T_R$.

11.3.6 Absorbing Set and Global Attractor

Let $\mathfrak{C}_\alpha$ be the subspace of $\mathfrak{C}([0,\infty), \mathbb{R}^d)$ consisting of functions f with $\|f\|_\alpha < \infty$. Then $(\mathfrak{C}_\alpha, \|\cdot\|_\alpha)$ is a Banach space and the operators $(T_t id_{x_0})(\cdot)$, $t \geq 0$, form a semi-group on $\mathfrak{C}_\alpha$.

Define the closed and bounded subset $\mathfrak{B}^*$ of $\mathfrak{C}_\alpha$ using the σ metric by

$$\mathfrak{B}^* := \left\{ \chi \in \mathfrak{C}_\alpha \ : \ \|\chi\|_\alpha \leq 2R^* + \frac{B_{R^*}^g}{\alpha \Gamma(\alpha)} =: \widehat{R}^* \right\}.$$

This set an absorbing set for the Caputo semi-group $(T_t id_{x_0})(\cdot)$ in $\mathfrak{C}_\alpha$ and absorbs bounded sets of constant initial data $\|id_{x_0}\|_\alpha \leq 2\|x_0\| \leq 2R$ in time $t \geq T_R$. Moreover, the semi-group is asymptotically compact. Thus a closed and bounded set of omega limit points in $\mathfrak{C}_\alpha$ exists and attracts all constant initial functions. In particular,

$$\mathfrak{A} = \overline{\bigcup_{\substack{D \subset \mathbb{R}^d \\ \text{bnded}}} \bigcap_{t \geq s} \bigcup_{s \geq 0} T_s(id_D)},$$

where $id_D := \left\{ id_{x_0} \in \mathfrak{C}_\alpha \ : \ x_0 \in D \right\}$.

Remark 11.1 The compactness of $\mathfrak{A}$ follows as in the asymptotic compactness proof. Let $\{\chi_n^*\}_{n \in \mathbb{N}}$ be a sequence in $\mathfrak{A}$. These functions are uniformly bounded and equi-Lipschitz continuous and hence have a convergent subsequence by the Arzelà-Ascoli theorem.

Remark 11.2 This justifies the earlier comment that Ω^* is the *observed* part of the attractor. In fact, given the skew-product like character of the Caputo semi-group it is the determining part of the attractor, since once it is known the other part of the attractor can be readily determined, in principle.

In the scalar case $\Omega_{\mathscr{B}^*}$ is, in fact, the global attractor of the corresponding ODE, see Chap. 10.

11.4   Endnotes

This chapter is based on Doan and Kloeden [3]. The monographs by Hale [4] and Kloeden and Rasmussen [2] provide background information on atttactors and dynamical systems. See Kloeden and Yang [5] for universes of attracted sets.

The results here have been extended to the non-autonomous skew-product case in Cui and Kloeden [6] and Doan and Kloeden [7].

References

1. Hale, J.K.: Asymptotic Behavior of Dissipative System. American Mathematical Society, Providence, Rhode Island (1988)
2. Kloeden, P.E., Rasmussen, M.: Nonautonomous Dynamical Systems. American Mathematical Society, Providence (2011)
3. Doan, T.S., Kloeden, P.E.: Attractors of Caputo semi-dynamical systems. Fract. Calc. Anal. Appl. **27**, 2305–2316 (2024). https://doi.org/10.1007/s13540-024-00324-x
4. Hale, J.K., Koçak, H.: Dynamics and Bifurcations. Springer-Verlag, New York (1991)
5. Kloeden, P.E., Yang, M.: Introduction to nonautonomous dynamical systems and their attractors. World Scientific Publishing Co., Inc, Singapore (2021)
6. Hongyong, C., Kloeden, P.E.: Skew-product attractors of non-autonomous Caputo fractional differential equations. Chaos **34** 083125 (2024). https://doi.org/10.1063/5.0214041
7. Doan, T.S., Kloeden, P.E.: Asymptotic behaviour of non-autonomous Caputo fractional differential equations with a one-sided dissipative vector field. Buletinul Academiei de Stinte a Republicii Moldova Matematica 104(1)–105(2), 44–52 (2024)